The unseen
GREAT WESTERN

Images from the Harry Collection at
STEAM - Museum of the **GWR**, Swindon

The Transport Treasury

The unseen Great Western

Front Cover: No 4986 *Aston Hall* standing on the Up main on the approach to Swindon. No date but the red headlamps indicate 1931 at the earliest (No 4986 emerged new from Swindon in January 1931.) The image is clearly taken to show the engine on the ATC* ramp - note these were not parallel with the actual running rails but laid at a slight angle so as not to wear a groove in the pick-up shoe - which is visible below the centre of the buffer beam. On the left are the Down sidings and to the right the Up goods line - beyond this is a siding for works purposes. We have spent some time studying this image under a glass (other similar views of engines at this location have we know been taken) whilst this particular image raises a few questions. Firstly the loco; look at the running plate and it appears to have some form of 'curtain' either side of the smokebox, whether this was a 'thin' spot on the negative we cannot be certain. A rear view of what is probably the same engine on the same occasion appears as Plate 109 (page 183) of David Smith's seminal tome 'GWR Signalling Practice'. Taking these two views together it is likely No 4986 was positioned as seen during a quiet period, a Sunday morning perhaps, with the shunting engine out of shot behind the photographer. Next to the fencing either side of the Up and Down main lines, simply put - why? Was this to stop workers taking a short cut and so bypassing the pedestrian tunnel to reach the works? It is hardly health and safety (should we not really say 'safety movement' feature?). Then to the ramp itself. Located on a main running line this is hardly likely to be a 'test' feature, such test ramps were normally fitted on the exit road to engine sheds as well as on various sidings in the works. Believing it to be an operational ramp, where is the signal which it referred to? A study of the diagrams for Rodbourne Lane and Swindon West fail to provide a definite answer. The conclusion; a case where an image which at first glance appears to be straightforward instead raises questions we are unable to answer. One other point concerns the loco lamps which from their dark surrounds are clearly red. Officially from 1903 the colour of loco lamps was altered to white but this change took some years to complete. Indeed in 1936 an instruction was issued to paint the cases of loco lamps white but even a decade later engines were running with a mixture of red and white bodied lamps. *Ref HC12A 1A2 (74064)*

Frontispiece: Looking out from the old Bristol Bath Road shed on 2 February 1932. The engine on the right is No 4068 *Llanthony Abbey* built as a member of the Star class in January 1923 subsequently rebuilt and renumbered, 5088 but retaining the same name (with the addition of 'Castle Class' on the nameplate) in March 1939. This engine had a total life of 39 years and was withdrawn in 1962. On the left is Bath Road Signal Box with the overall roof of Bristol Temple Meads station in the background. This signal box which, as the name suggests, controlled the entry and exit to and from the shed had opened in 1898 and closed on 12 December 1933, its size was 17' x 12' x 8' to the operating floor and we know it had a 14 lever frame. The whole station area was then in the process of remodelling and extending, this in connection with the 1929 Government Loans Act, an act intended to assist with new infrastructure projects and the relief of the then chronic unemployment situation. In addition to improving the passenger station Bath Road depot was completely rebuilt whilst signalling in the area was overhauled, mechanical signalling giving way to colour lights. As a stop-gap a temporary 'Bath Road Signal Box' was erected on a different site and opened on 10 December 1933. This time we know the number of levers was 80, but not the size of the structure. It lasted until 26 May 1935 when entry and exit from the shed was fully switched to 'Bristol (Temple Meads) Loco Yard' box. Note also the roof detail of the vans alongside the engine, these are from the depot breakdown train. Finally, on the right the incline leads up to the coaling stage. *Ref HE2 006 (85325)*

Rear Cover: To me it matters not if the odd view may perhaps have been seen before and this is one that may well fit into that category. The 'Is it Safe' movement is a topic that may one day be fully researched and deserves to be written about. Meanwhile alarmed at the number of injuries to staff of all grades the GWR posed a series of views covering almost every department. Here we have the porter riding down a slope on a 4-wheel trolley. Clearly something had happened - somewhere - to warrant its inclusion. The company also issued 'Is it Safe' tokens intended to be kept in a pocket to remind the holder of the dangers he might be exposed to at work, and with railway workshops and railway operation that was a considerable number on a daily basis. *HS2 001 (53870)*

* The initials ATC are the common abbreviation for the GWR's 'Automatic Train Control' *NOT* to be confused with the later BR AWS (Automatic Warning System). ATC involved placing a ramp to the rear of the distant signal which would be energised only if the distant signal was showing 'off' (at clear). This ramp made contact with a shoe on the engine which, if the ramp was energised, would ring a bell in the cab. However, if the signal were 'on' (at danger) no current was applied to the ramp and on the shoe making contact with the ramp so a warning horn sounded in the cab which this time, if not acknowledged by the driver, would automatically apply the brakes and bring the train to a stand even with steam on. In this respect it was 'fail safe'. ATC was a major factor in the enviable safely record of the GWR.

Copies of many of the images within *The Unseen Great Western* are available for purchase/download.

Please contact www.steampicturelibrary.com

© Images: STEAM Museum Swindon, text: Kevin Robertson, design: The Transport Treasury 2025

ISBN 978-1-915281-20-3

First Published in 2025 by Transport Treasury Publishing Ltd., 6 Highworth Close, High Wycombe HP13 7PJ

https://www.transporttreasury.com

Printed by Short Run Press, Bittern Road, Sowton Industrial Estate, Exeter, EX2 7LW.

The respective copyright holders hereby give notice that all rights to this work are reserved. Aside from brief passages for the purpose of review, no part of this work may be reproduced, copied by electronic or other means, or otherwise stored in any information storage and retrieval system without written permission from the Publisher. This includes the illustrations herein which shall remain the copyright of the respective copyright holder.

INTRODUCTION

This book is aimed at a specific audience, the Great Western enthusiast. It is to be hoped it may perhaps find its way on to shelves of some others but I will not be so naive to think it will appeal to many who follow 'other' railways.

Sticking my own neck out, I have always been primarily a GW man; albeit with a leaning towards certain aspects of other systems and designs as and when the mood takes.

In consequence I have devoured numerous tomes on my pet subject(s) with the book shelves at home groaning and my respective wives given up passing comment when yet another book arrived. (Note; despite the deliberate use of the plural 'wives' I assure all it has only ever been one at a time.)

Being serious now, in consequence of amassing a collection of books running well in to four figures plus, and having been involved for many years both in publishing and the retail arm of railway books I think I can say that literally thousands of images have passed before me. Even so it is still a privilege to see something new and there certainly is still plenty to be uncovered.

Coming to work at STEAM as a volunteer in the library I was made aware of an archive I had personally never previously heard of, the 'Harry' collection. How, what, why 'Harry' is a story in itself and I am most grateful to Elaine Arthurs, Collections Officer at STEAM for describing its journey on page 104.

From the perspective of a photographic researcher the collection is a veritable gold mine, countless images compiled from a variety of sources including the official GWR/WR collections and which to me are new or show items from a different perspective.

Because many are, to me at least new, the choice of title is therefore obvious, however I am fully aware there will be some that are perhaps familiar to readers, whilst others may look similar but are perhaps seen from a different angle.

Of the official images, I suspect a few are 'office' or 'file copies', and it is very likely the negatives for these simply do not exist any more.

I should add this is also a book that has been some time in the gestation. The reason simply that I wanted to get it as right as possible. I will also state I do not for one moment profess to be an expert, instead one who has a working knowledge in various subjects and perhaps a greater or lesser appreciation in others. Asking questions of and securing answers and opinions from others is what takes the time; but can also be so rewarding.

In setting out this work I also wanted it to be a book of some surprises, hence I have deliberately ignored the convention of introducing views by time-scale, subject, or geography. Instead it is presented in what I hope is the simpler order or 'images that work well with each other'.

The 100+ views that follow represent only the smallest fraction of the 27,000 plus images and other material within 'Harry' and for the present concentrates (not exclusively) on rolling stock and locomotives, for variety we have also strayed into the early BR era on a few occasions. Certainly not all within 'Harry' is suitable for reproduction but it is to be hoped what is presented in the pages that follow is a flavour and could lead to more in the future.

Kevin Robertson, Lambourn 2025.

ACKNOWLEDGEMENTS

Without the help of others, the information on the following pages could not have been compiled. Elaine Arthurs, Mike Barnsley, Jeremy Clements, David Collins, Amyas Crump, Steve Gregory, Andy Malthouse, Bob Meanley, Allan Pym, Andrew Royle, Dennis Troughton, Marlene Wheeler plus all the other library volunteers who have been valuable allies, I express my gratitude to them all.

In addition reference is made in the captions to specific books written by various individuals most of which will already be familiar to followers of the Great Western Railway. The intention of any author is to enhance knowledge already available and this is indeed the aim of this work but to also highlight the vast archive that exists at STEAM running to in excess of 100,000 items.

Note: 'Artificial intelligence' has *NOT* been used in the compilation of this work.

The unseen Great Western

An unusual set of five vehicles were the Bullion vans built by the GWR under three separate Lot numbers and to Diagrams M16 and M17. First to appear were two vehicles Nos 791 and 792, to Lot 996 on 11 April 1903. These set a new standard and understandably when considering the load they would carry had bodies of all steel construction. They were non-corridor and could carry a load of 16 tons distributed. A pair of heavy plate-frame Dean bogies was fitted. Interestingly they also had eyelets along the roof line for communication cord purposes which were retained throughout their lives. Two further vehicles this time to Lot 1139, Diagram M17 were built in 1908, the difference now that they ran on pairs of 9' American bogies (these were the type of bogie having both vertical helical springs and horizontal leaf springs), this latter pair, Nos 819 and 820 had a wheelbase of 33'. Again they were fitted with roof eyelets. A final vehicle No 878, again to M17, Lot 1220 was built in 1908. It was identical to the 1908 pairing excepting it did not possess the roof eyelets. Dimensionally all were 36' x 8' overall with a maximum internal height of 7'. Initially their sole purpose was the safe transfer of bullion between Paddington and Plymouth. Of principal concern was the security of goods in transit and apart from the type of construction they also only had access doors on one side; the other side as seen here blank, (note Russell reports '...some of the vehicles had doors on one side only...', and by deduction he means 819, 820 and 878). Clearly then there must have been some pre-planning to ensure that at Paddington a train containing one of more vans was routed into the correct platform! Even so the whole operation of these vehicles raises a number of interesting questions. Firstly did the transfer of gold bullion only become important at the start of the 20th century? If it was taking place earlier, does this mean there was concern over security or had an incident occurred? We know that bullion van(s) were often included in the formation of an Ocean-Liner service where they were always marshalled immediately behind the tender, so does this mean there was a human security presence either within the vehicle or on the actual train? As built, the vans would have been out-shopped in the crimson lake livery of the period, including the final vehicle No 878 (see Russell Part 2 page 52).

Realistically all five would spend most of their lives in standard chocolate and cream. Shown on the '88D' model website No 878 is seen in colour at Snow Hill and still retaining these colours. It also proves the point the vehicles were not restricted solely to the Plymouth service - were they in fact used for any other purpose other than bullion - although possibly No 878 was then carrying a similarly valuable load in the Midlands. Branding details are known for three of the vehicles, Nos 819 and 820 'Return to Plymouth' and No 878 'Return to Paddington'. One change is known with No 820 in 1938 when it now instead read 'Return to Paddington'. Branding for the first two vehicles is not reported.

At least one, No 792, was repainted in BR 'plum and spilt milk' and the same vehicle was also noted later in all over maroon. For specialist vehicles three at least had a long life, No 878 withdrawn in 1959 with Nos 819 and 820 lasting until 1967. The fate of the first two vehicles, Nos 791 and 792 is not reported. In this undated image two vans Nos 819 and 820 (identified by the bogies and roof eyelets - assuming no bogie swaps had taken between this pair and the first two vehicles), are seen in Platform 4 at Paddington and possibly in different liveries; note too the rear of the lorries drawn up on the platform which may well be being used for unloading. The vacuum hose has also been uncoupled from the leading van. Away from the vehicles loading gauges still in position over platforms 3 and 4. *HC4 088 (6261)*

The unseen Great Western

Tunnel inspection / gauging vehicle at Swindon converted from a Dean 40' van. Five such conversions took place between 1949 and 1952 and which were spread around the Divisions. Russell (Coaches Appendix Vol 2 Page 260) illustrates a similar vehicle converted from a 40' passenger brake van although in that example the corridor connection was retained. See also the GWSG publication on Dean 40' Vans. Unfortunately there is no detail as to specific identity of the vehicle illustrated. In addition to these types of vehicles converted from former passenger / parcels stock some 'Toads' were similarly modified; the latter it seems were mainly used in the Severn Tunnel and in all cases what we would now term 'recycled' from their former use (before such a word became fashionable). Tunnel inspections were a vital part of infrastructure maintenance and could take the form of a simple walk along the line observing from ground level or a more detailed view from the top of a converted item of rolling stock. This would take place on average annually. Most tunnels would have (several) courses of brick lining and it was important for these to be observed to check for wastage, spalling, water ingress and even bulging. Track maintenance within a tunnel was generally left to the local ganger who would be paid a 'tunnel allowance' as well as be issued with protective clothing if the tunnel on his length was longer than 440 yards. (At Winchester where the tunnel was on a slight curve the length was officially set at 439 yards, however the local ganger petitioned an appeal and the tunnel was measured again, this time on the longer side and came out at 441 yards - the men got their allowance.) Tunnel conditions can also require much higher levels of track maintenance as indeed occurs in the Severn Tunnel even today. In addition to the vehicle seen here there were also vehicles equipped with what were in effect an arc of 'prongs' to check concentricity especially when any track renewals had taken place, this would ensure the loading gauge was maintained. (Later on the WR, at least three former passenger vehicles were known to be used in connection with tunnel inspections. The first was DW150341, a former Collett corridor brake second 9605 based at Cardiff Canton. Next was DW150173, again a former Collett BSK, converted in December 1958 again as a tunnel inspection staff coach and based at Radyr. Finally another Collett BSK, the former No 5785 renumbered and converted as DW150340 in May 1963.) HC1 022 (48955)

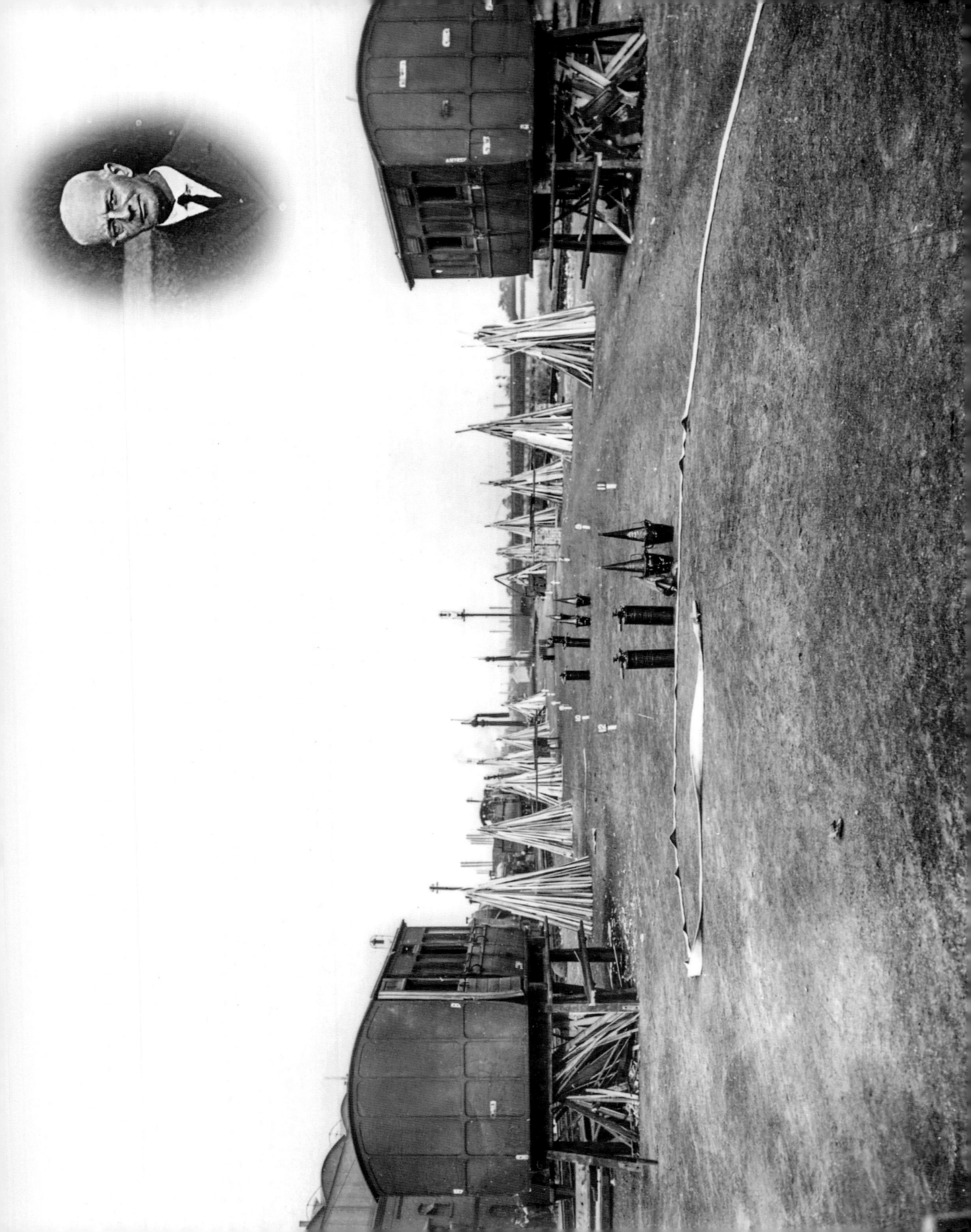

The unseen Great Western

Opposite: Pyrotechnics at Old Oak Common circa 1910. Coming across this view with written on the reverse 'Old Oak Common Fire Competition' immediately raised the question might this have been some form of social event...? Further digging and cross referencing instead revealed it to be a test of the efficiency of different types of fire extinguisher, hence the various items in the centre and an emergency hose on the ground in case of need. Thus far from any type of unofficial gathering it was a serious experiment under the direction and control of a Cecil Taylor Cuss of Swindon. Mr Cuss was a member of the Drawing Office staff where he had been appointed on 4 October 1898; it should be mentioned there were several other Swindon employees with the same name. From 8 September 1902 he was put on the 'staff' list by Churchward and meaning he was now salaried rather than weekly paid. (Some salaried staff received their remuneration at six-monthly intervals.) Mr Cuss was a man who seemed to be charged with undertaking a number of slightly off-beat tasks. He was a regular contributor to the Swindon Engineering Society and as a example in 1903 presented a paper to the Junior Engineering Society on sewage disposal and the following year a paper on Locomotive Water Supplies. The inset cameo depicting him (with other participants) came in consequence of winning a rifle club competition. Further research established a total of 12 images of the Old Oak conflagration were reported, obviously we would like to have seen others. The above view has the date '27 June 191? - with the last year digit missing. Without further information it is impossible to draw definite conclusions but from the view itself it appears the 'wigwam' type structures will be lit first and with the final test on the coach bodies; possibly the same vehicle literally sawn in half - might it even have been of broad gauge origin? The land here was later given over to carriage sidings. Although retired by the time of WW2 Mr Cuss returned to Swindon to supervise the erection of the BSE (steaming) shed at this time. *HA2 012 (69845)*

Above: Another view which has taxed the grey cells. According to the note written on the rear of this print the subject is 'WD Ambulance Coaches'? After some thought, a simple conclusion - wrong! Firstly the vehicles seen are hardly suitable for ambulance use, most seen being open or compartment stock. Ambulance trains tended to use vans as 'ward cars' which could accommodate stretchers fixed parallel with the sides. In addition the stock is simply too modern - well most. After deliberation the conclusion is this is a 'static office' view from the roof of the carriage paint shop at Swindon. Turning initially to the vehicles we have Collett coaches, a Mulitbar Toplight' No 3954 (Third Class to Diagram C35 built in 1919 Lot No 1256) on the far right and a catering car on the far left. Possibly there is also a stores van on the siding behind the Toplight. Apart from the Toplight the various rakes have one thing in common which is access to an external steam heat connection from the works. Several other items come to note starting with the pair of vehicles on the right (line five). Notice here the makeshift pole vertically at the end, on top of which is a cable - power or telephone communication? There is also a set of access steps to this coach. Whilst no obvious steps / ladders are seen for the other rakes that is not to say they may not have been positioned elsewhere. Counting from the left, lines 2, 3 and 4 are confirmed as a rake of open or composite vehicles by the position of the roof water tanks, all dating from the Collett era. An end corridor connection is missing from the vehicle on Line 3. Again with Line 3 notice the clean(er) roofs of two of the vehicles. Away from the railway, observe the two square turrets each with a scaffolding surround on the roof of the building on the skyline. These were anti-aircraft guns positioned on top of the premises of Messrs Compton Sons & Web. This company made various items of uniform for the armed forces and civilian services. *HC1/100 (53066)*

(Note: The reader should not start to believe that the records held with 'Harry' are all similarly incorrect. Simply that in some cases, indeed as with collections elsewhere, no doubt a well-intended individual has at some time scribbled on the reverse of the print what they believed to be correct - the official registers are not always accurate either. With in excess of 100,000 images in the STEAM archive it will take some time to go through and check all the information listed.)

The unseen Great Western

Above: This particular image was chosen even though there was no information as to where or when - or what (but then we like a bit of detective work... .) The only occurrence found that may be connected was an accident that occurred in fog at Northolt Junction on 22 January 1934 when the 2.34pm LNER train from Marylebone to Ruislip ran into the rear of the 2.35pm GWR service from Paddington to Oxford via Thame). Three coaches of the GWR train were damaged but the vehicles were sparsely occupied and there were no serious injuries. The GWR vehicles were later moved to a siding (not stated where). The aftermath shows the GWR train to have consisted of suburban articulated stock, the bodies of at least two vehicles seemingly having shifted on their common bogie. Despite the lack of information the view was chosen for the coach detail provided, some clearing up of the bodywork having already taken place, allowing the vehicles to be moved. We can assume there was some action taking place nearby as witness the onlookers at the top of the embankment. The alternative scenario is a wartime incident. The label on the extreme right 'West.....Pad....' could be more indicative of a suburban working but again can we take the label to be accurate for the service? We also know the articulated suburban stock was known to work as far out as Oxford. The coaches seen here were also rebuilt as the articulated sets survived until 1957-60. *HA1 087 (72012)*

Opposite top: Staff accidents were an unfortunate aspect of railway operation, a fact identified years before with the start of the GWR 'Safety Movement'. As time progressed so ever more emphasis was placed on safety with the official photographer becoming involved to record locations of accidents especially it seems around Swindon Works - and where there was a photographic studio based. This illustration was taken to record 'Accident to Shunter Wyse on turntable, 7 January 1953'. We are not told the circumstances of the accident nor the fate of Mr Wyse. Certainly at one time there would have been a file relating to the event but this was likely lost years earlier. Having the date is useful as with the letters 'GWR' still readable on the tank side the temptation might otherwise have been to refer to a pre-1947 occurrence although that could be immediately countered with the corner of the smokebox numberplate just visible on the 4-6-0 behind ('Manor' class perhaps?). The works turntable still exists in 2025, albeit fenced off and next to one of the car parks that occupy the site. Sadly No 1709 succumbed many years ago. Built with a saddle tank in 1891 as a member of the '1854' class it was rebuilt with pannier tanks in 1913. Officially withdrawn from Newport (Pill) on 7 November 1950, it clearly survived for some time further as a works shunter. *HL052E 072 (57862)*

Opposite bottom: Aftermath of the Appleford accident that occurred in the early hours of 13 November 1942. This involved an unfitted freight train of 72 wagons and brake van with a total weight of 770 tons, hauled by No 2975 *Lord Palmer*, just fresh from overhaul, which overran the outlet signal at the end of the Down Appleford Goods Loop causing the following wagons to pile up behind and in so doing foul the Down main line. As this was taking place it was indeed unfortunate that passing on the adjacent Down main line was No 4086 *Builth Castle* hauling 11 bogies forming the 12.00 midnight Paddington to Birkenhead express. The formation of this train was five vans leading, three passenger coaches and a further two vans. The engine and first two vehicles of the Birkenhead train managed to pass the scene without incident, not so the remainder and it was indeed fortunate that of the 200 passengers on board the train just two persons died, one passenger and an on-duty porter. Considering the impact it was also lucky that only a small number of passengers were injured. Not so the engine crew of No 2975 who were both killed. Full details of the accident may be found at https://www.railwaysarchive.co.uk/documents/MoT_Appleford1942.pdf . No 2975 did not survive long after the accident and was withdrawn in 1944. *HA1 019*

The unseen Great Western

The unseen Great Western

The unseen Great Western

Opposite top: A local view to Swindon, referred to as the 'Transfer Bridges' and probably still recognisable to many. This image shows what was known as the 'Transfer Bridge' looking north with Station Road on the left (the latter now one way) and the entrance to the station goods yard on the right. The road dips to give some clearance under the structure but it remains limited even today although it has been levelled to prevent flooding in the dip. No date is given but we might suspect at the latest 1950s and possibly a bit before. Compare also with any such scene today and the lack of vehicles is perhaps the most obvious, just one car parked in the distance - many of the buildings in the vicinity of the car demolished long ago for various new road schemes. Local goods traffic at Swindon also ceased years past with the plethora of signs part inviting traffic and at the same time warning of the (dire) consequences of trespass and loitering. Standard railway 'spear' fencing is present; some still remains today in odd places on the network - the 'Health & Safety Police' have not been on a prowl everywhere - yet. Otherwise enjoy a name from a bygone era 'Corona', for example which brand ceased as recently as the late 1990s. To conclude we have the shadow of the photographer just visible on the roadway and a festooned telegraph pole in the left background. *HB2 004 (71246)*

Opposite bottom: Now we have the Wootton Bassett Road bridge, looking north and stated as 'before reconstruction'. Evidence of work required is clear from the supports to the retaining wall. The view is interesting as through the bridge it also shows the MSWJ line running left to right; Swindon Town towards Cirencester. Possibly the small group of men, inspectors probably as witness the presence of the 'trilbys' are contemplating the work required in the area. Near to where they are standing is the clear outline of a 'Toad' but it of course impossible to tell it the train is stationary. What we can be more certain of is the 'Down main home' signal is cleared for a train on the main line, the stop signal on the lower doll and associated fixed distant signal applying to the left hand junction to the MSWJ line at Rushey Platt Junction a short distance to the left. A BR view as this a tubular signal post. As time passed so this road was widened and the MSWJ bridge would eventually be removed completely along with the adjacent embankment to allow for alternative development. Strange to think too the works sidings came almost as far as the land on the right between the two bridges. The firm of Weston Cider were established in 1878 and still in business today. Victoria Garage however is defunct as indeed is the 'Riley' brand of motor car - and four digit telephone numbers. *HB2 008 (71250)*

Above: The River Severn bridge near Buttington Junction north towards Oswestry. Opened in 1860 the station at Buttington was also the junction for a line through Westbury (Salop) to Shrewsbury. Standing in the 'four foot' the local ganger has his key hammer over his shoulder. As he walked the length daily he would look for loose or missing keys; years of experience meaning he could tighten a loose key with just one swing of the heavy hammer. We are not told the reason for the view. To record the man but no name is given, or simply to record what is a pastoral scene, notice the 'Beware of Trains' sign on the right hand side, probably at the top of a path from the river. The man is in the traditional 'uniform' for his grade, not formally issued but men in this role seemed to have worked out for themselves the most appropriate garb. He would be part of a gang consisting a sub-ganger and various lengthmen, the remainder of his team no doubt engaged on a task which could be anything from grass cutting to packing ballast or even changing a rail. Notice the track circuit bonds across the joint on the right hand rail and the tell-tale 'diamond' on the signal in the background. *HB1 088 (71229)*

The unseen Great Western

There are several files of station views within the collection and that does not include the 'station' files separate from 'Harry'. Be aware however, not every station is featured in either archive and even if a particular location is included it may be a 'standard' view - meaning seemingly everyone who ever went to the location took the same basic image! Here we have Arenig on the line from Bala Junction to Blaenau Festiniog. The station was one of five crossing places on this 25½ mile single line, the other places where passenger trains might cross being Bala, Trawsfynydd, Festiniog and Manod. The station served a thinly populated area but lack of passenger traffic was made up for by the nearby granite quarry which provided ballast for much of the GWR. Operation of this railway in its early years would have been interesting. Opened as a private concern in 1882 at first there was no direct connection beyond Festiniog and passengers wishing to continue their journey had to board a narrow gauge train to Blaenau. This arrangement continued until the following year when the route north of Festiniog was converted to standard gauge although narrow gauge trains continued to operate until September 1883. Mixed gauge working did not just occur in the south! The independent company that had promoted and built the railway was eventually absorbed into the GWR in 1910. The view is looking towards Blaneau with the main building and signal box on this platform. The view is undated and appears to show a passenger working with tail lamp affixed; perhaps a 'short working' which has arrived and the engine will then return the train whence it came. Inside keyed track is also visible. Just visible is the signal box which dated from 1882 and lasted until closure. The original configuration is not known but on 18 October 1905 is was reported to contain a 25 lever frame at the larger 5¼" centres with stud locking. Until about 1940 a 2' gauge narrow gauge railway had brought stone to the station but this was then replaced by a conveyor where it entered a crusher close to the station. The actual stone was also in close proximity to the station and before any blasting took place the railway had to confirm the timing was acceptable. Arenig was closed in 1960, single figure passenger numbers per day meaning the decision was inevitable. Freight between Bala and Blaenau continued until 1961 although part of the track bed was subsequently deliberately flooded to form a reservoir for Liverpool Corporation. *HSA3 008 (100800)*

The unseen Great Western

The important location of Barmouth Junction on the former Cambrian main line. Here the railway forms a triangle with the Cambrian branch to Dolgelley and which in turn forms an end-on junction with the Great Western line from Bala Junction. This image is taken from the north end of the triangle with the main line from Pwllheli behind the camera and the main line to Dovey Junction to the right. That to the left is the branch to Dolgelley and eventually Ruabon. Today it is only the Pwllheli line that remains open. As with many locations the railway here opened in stages and with facilities that grew as time progressed. First was the line from the south to Penmaenpool (on the way to Dolgelley) in 1865 this later became known as the loop. The main line, but as a this stage at a terminus, followed in June 1867. For a month trains could not cross the bridge over the estuary at Barmouth and passengers and goods were conveyed by horse drawn transport until the main and branch lines opened to passengers on 3 July 1867 and goods on 10 October 1867. The land hereabouts was mainly consisting of marsh and at the time of opening much of the railway had to be floated on brushwood whilst the station needed piling. In respect of access it was similar to that at Dovey Junction with poor access from the nearest road via a 'grace and favour' arrangement or else by trespass. The view is of the North signal box, at one time known as the station box. This worked to South box so far as the main line was concerned, and to East box for the line to Dolgelley. The loop making the third side of triangle, and on which there was never a passenger platform, allows train to run direct from the direction of Dovey Junction to Dolgelley without the need for the engine to run round or interfere with traffic already in the station. Considering that today the station is still open as an unstaffed halt although without even so much as a passenger shelter - and it can certainly blow at times. The stopping place was also renamed Morfa Mawddach in 1960 to avoid confusion with the similarly named Barmouth station. Under Great Western ownership the location was an important interchange and for a time there was also a Camping Coach located here. Back in time there were drainage ditches alongside the loop and a 4' gauge tramway to some nearby quarries although this had ceased use by 1911. Apart from the expected marshland within the triangle there were also four railway cottages each with their respective garden, a pigsty and a cattle shed. The signal box seen here was timber and opened on 1 May 1892. (It lasted under the Great Western until 1931 when it was replaced.) Inside was a frame of 32 levers. In this view the signals are all of Cambrian (contractors) origin. On the outside of the left hand running rail there is a fouling bar, an unusual but not unique position for such a safety device. (Most fouling bars were later replaced with the simpler electric track circuit.) The idea was that to lock the facing points the bar would rise and fall, impossible to do if a vehicle were immediately in the vicinity of the turnout. The round point rodding may also be noted, once commonplace, it is now a rare feature having been replaced with inverted channel. As passengers of necessity might have to wait for a connection at the station the usual facilities included Ladies and Gentleman's conveniences were provided along with a refreshment room. There was no goods yard but there was a bay platform for trains to Dolgelley and in the early years a private siding ('Andrew's) for transshipment of goods from what was Soloman Andrew's tramway. *HSB1 055 (49159)*

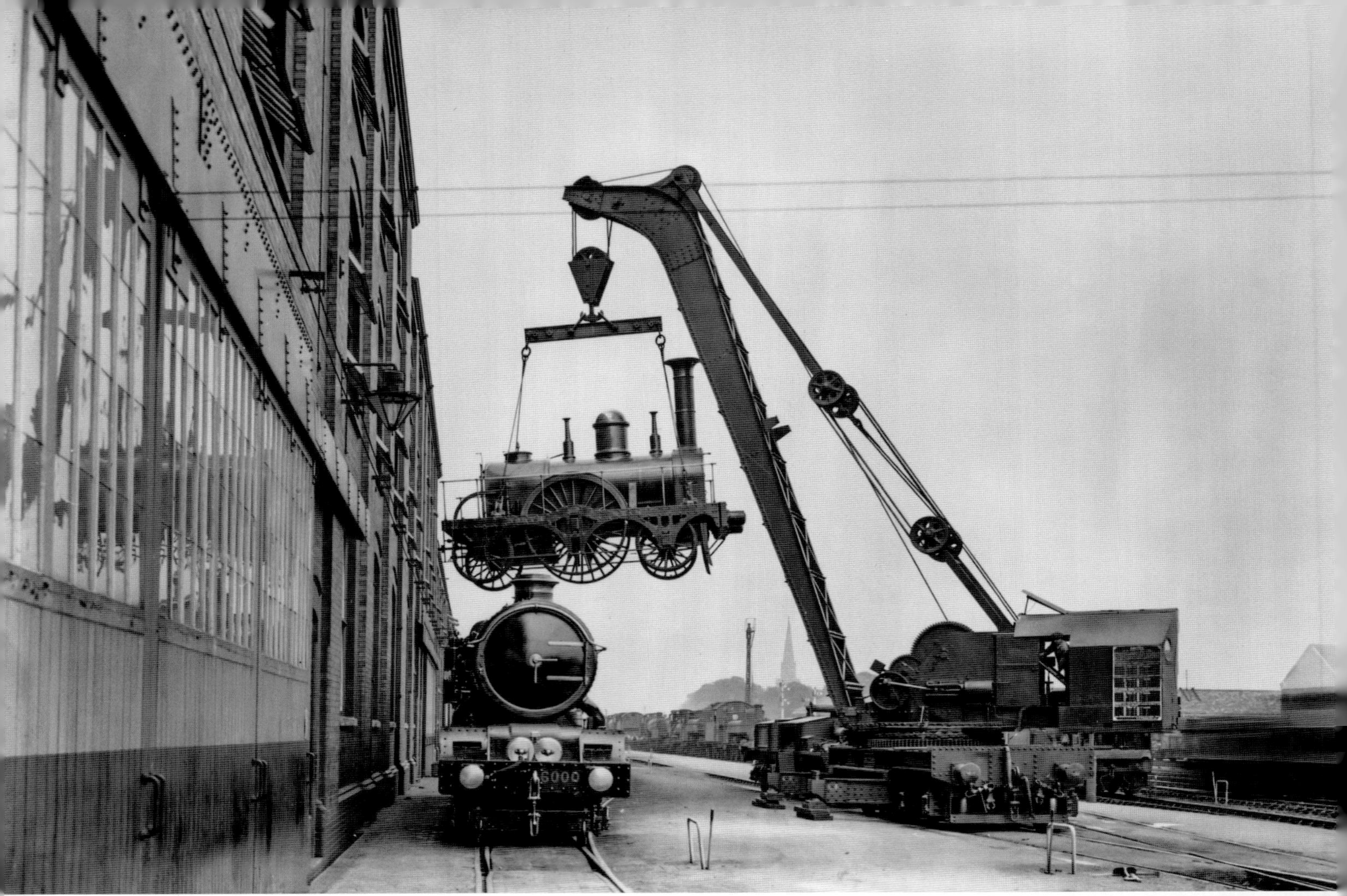
6000

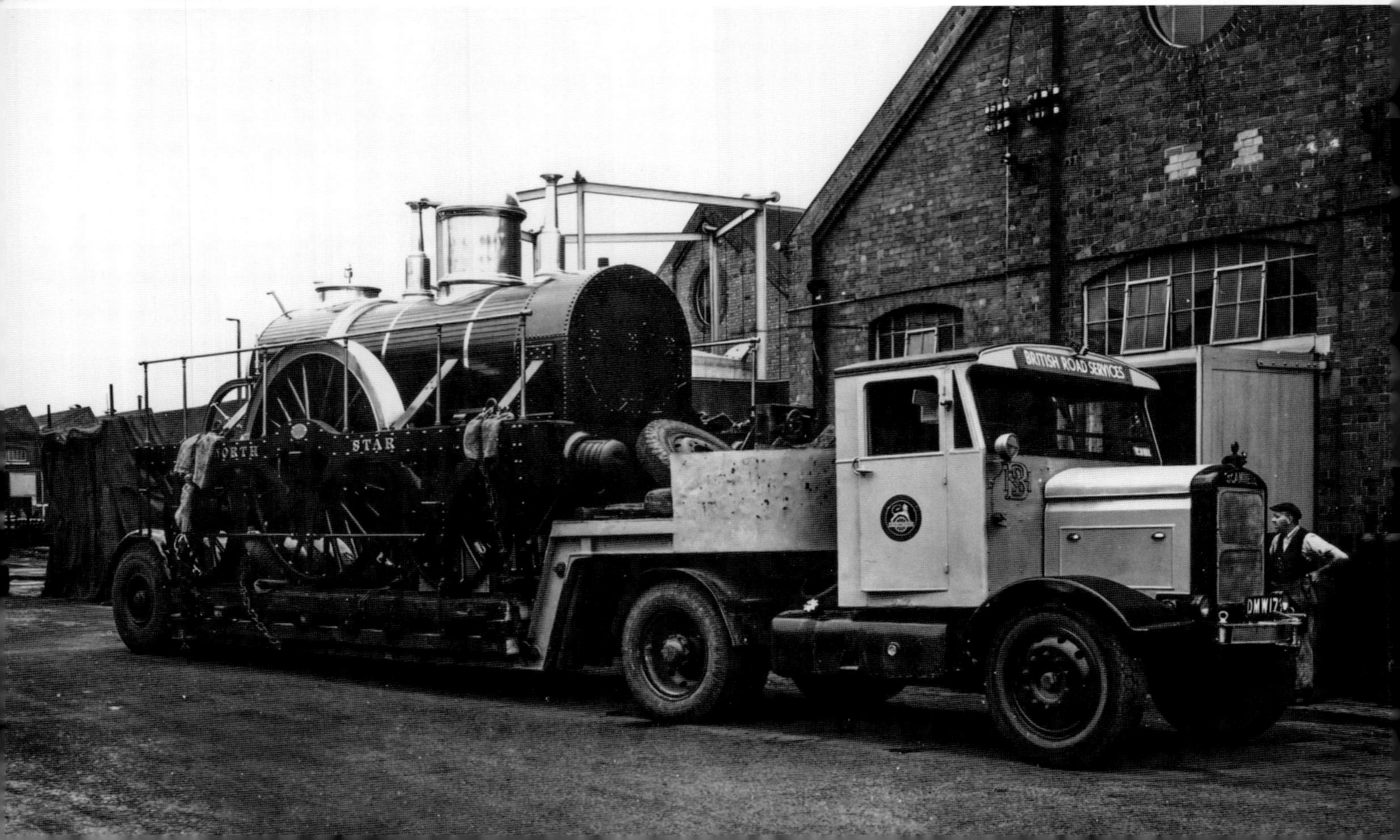
BRITISH ROAD SERVICES
ORTH
STAR

The unseen Great Western

Opposite top: Publicity was very much at the forefront of the GWR especially when it came to prowess, progress and pastoral scenes. Here we have an example of the first two with the replica *North Star* being hoisted above a brand new No 6000 *King George V* at Swindon. The date is shown as 28 June 1927 and so taken just a few days after completion of No 6000 and before its despatch to America in August 1927. (No bell on the front framing - yet, although note the Westinghouse pump on the side of the smokebox. The tender that accompanied No 6000 was also fitted with a buck-eye coupling). Clearly they were proud of their latest engine and what better way to showcase this with a comparison of 90 years of locomotive development. No 6000 was the doyen of the 'King' class of which 30 would be built (more precisely 31 after No 6007 was written off following the 1936 Shrivenham accident). The original *North Star* holds the distinction of being the prototype of the GWR's first standard locomotive class and the first successful type for the company after Brunel's nineteen 'oddities' although this came about only because its originally intended customer, the 5' 6" New Orleans and Carrollton Railway was unable to conclude the purchase; this despite the engine having been shipped to America and in consequence returned. Re-gauged by the builder Robert Stephenson & Co, and with the driving wheel diameter also increased from 6' 6" to 7' 0' it arrived by barge at Maidenhead Bridge station on 28 November 1837. The engine worked until 1871 having been rebuilt over the ensuing 34 years with larger cylinders and a lengthened wheelbase. When withdrawn it was retained in its final condition and preserved at Swindon until the following minute appeared in the GWR Locomotive, Carriage & Stores Committee Minutes of 22 July 1903. '19. Mr Churchward reported that the old Broad Gauge Engines *Lord of the Isles* and *North Star* had for many years been stored in a shed at Swindon and that the space occupied by them is much needed. Having regard to the interest attaching to the two Engines, the Committee consider that they might, with advantage, be offered to the South Kensington Museum & they agreed to recommend that an endeavour be made to dispose of them in this manner.' Then on 21 December 1905, '8. Referring to minute No.19 of the 22 of July 1903 the Locomotive Superintendent reported that the old broad gauge Engines *North Star* and *Lord of the Isles* which occupy much valuable space in the shops at Swindon have been offered to several institutions without success and upon his recommendation the committee approved of the same being broken up.' According to the 'devboats.co.uk website 'Neither locomotive was completely lost. Various parts of *North Star* survived in dark corners and even private houses, and according to the GWR magazine wheels, crank axles, eccentrics and even one leather buffer from the original North Star were used in the reconstruction. One half of *Lord of the Isles* motion was used to create an instructional model for a Swindon school, and this is now with the Steam Museum at Swindon, as are her driving wheels, name plate and works plate. It would appear the lack of a locomotive from early days continued to irk some even years later and a non-working replica was produced in 1923 of the engine as it had been at the time of withdrawal. Evidently there was some original parts in the 1925 replica shown at the Stockton & Darlington railway celebrations and in 1927 this replica was shipped across to America with No 6000. Both machines returned and *North Star* and *King George V* may at the time of writing (2025) both be viewed at STEAM. *HL14A 010 (53276)*

Opposite bottom: Following its return from display *North Star* was placed on a plinth inside Swindon works where it remained for several decades. It was removed when the works closed and for a time was on display at York before returning to Swindon. Here we see the replica loaded on to a 1930s British Road Services Scammell tractor unit with trailer. Unfortunately no date is given but it is very likely this was in the 1950s and when for a time it had been on display outside Swindon Town Hall. *HL01 007*

Above: From images the railway was proud of, to one they would rather keep quiet about. No date or circumstances but we can say it is a Ransomes & Rapier 45T crane dating from 1940 and one of four supplied to the GWR at the behest of the government, the cost was £9,425 per machine. From the attire, we might suggest 1950s. Clearly what has occurred is when attempting to make a lift the ground alongside the track has given way and the crane toppled. The rails look none too stable either. There is obvious damage to the base of the jib but without detail as to the actual crane we cannot say whether it survived this mishap. The (G)WR were not alone in crane mishaps with all four railway companies / regions the victim at some time or other. *HA1 005 (64895)*

15

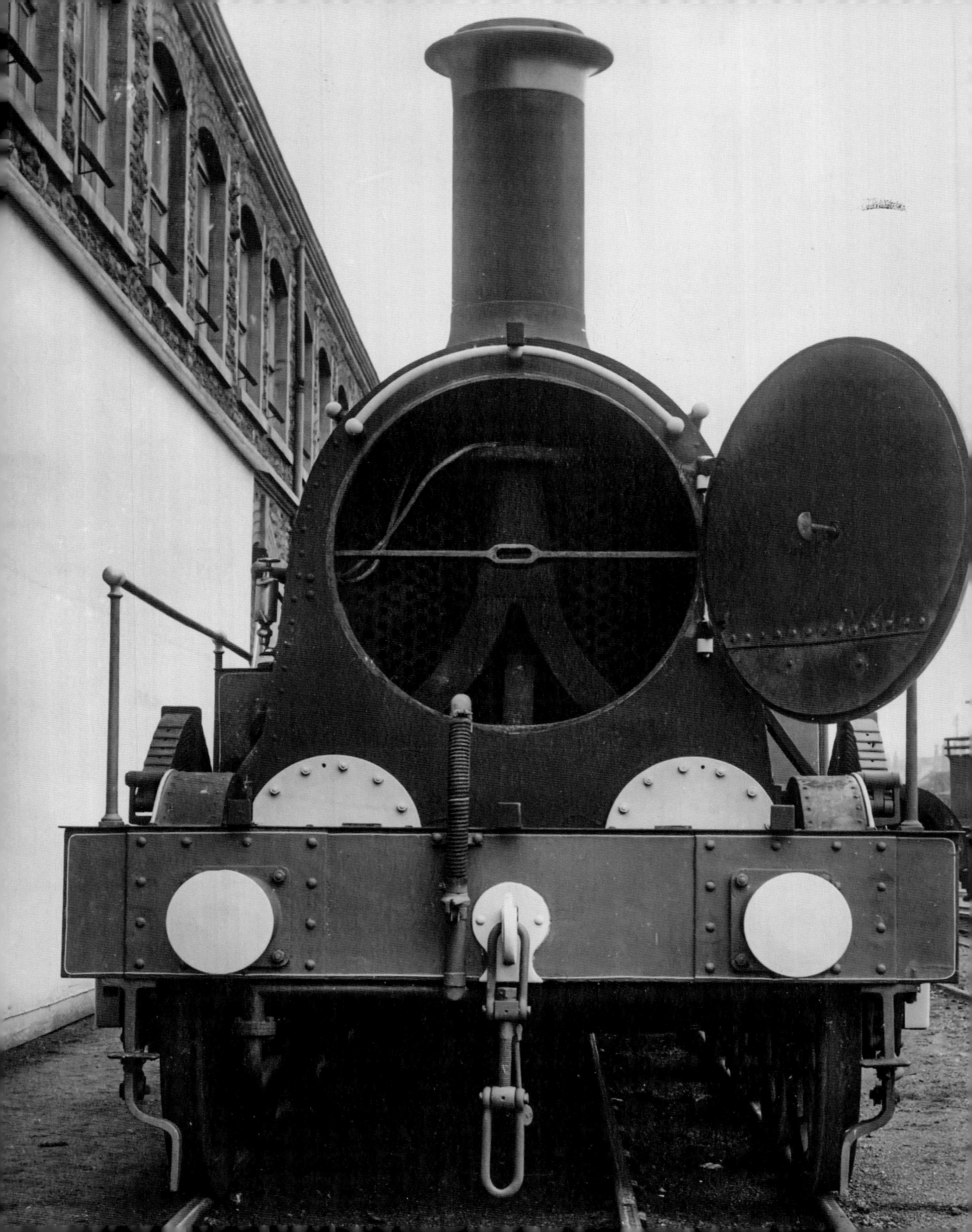

The unseen Great Western

Two splendid views of the Broad Gauge 4-2-2 *Bulkeley*, one of 24 engines of the type built at Swindon between 1871 and 1888 and officially known as the 'Iron Duke' or 'Rover' class. Officially referred to as renewals or rebuilds this is a slight misnomer as it was only the three engines of the type that were completed in 1871 that may have contained some parts from previous engines. *Bulkeley* dated from July 1880 but owing to the demise of the Broad Gauge ceased work at the end of May 1892. The type were better known as 'Rovers' whilst successive changes over the years saw an alteration in the physical size of the boiler, the number and diameter of the tubes, and in consequence the heating surface and grate area. *Bulkeley* is seen here in what was probably its final condition with an 'enclosed' cab, The first engines from 1871 had only a front weather board as protection for the crews but this was changed to an iron-roofed cab for those built from 1873. In service this was found to rattle badly and a wooden roof was substituted in 1876 and became standard thereafter. Tenders carried 3,000 gallons of water. That champion of the Broad Gauge, the RCTS book series gives detail of one of the class *Lightning* achieving 81 mph down Wellington bank in charge of five coaches. Faster speeds though, unless assisted by favourable gradients, invariably meant an increase in fuel consumption and economy was certainly the word for several years following the banking crash of Overend, Gurney & Co. in 1866. *Bulkeley* may or may not have achieved similar during its life although one actual run is recorded in the RCTS tome courtesy of the SLS Journal for March 1941. This took place on 10 December 1868 and involved a run from Paddington to Swindon with seven coaches when the engine was just 15 seconds over the scheduled 87 minutes for the 77 miles. The average speed was 53mph and the maximum noted as 60mph.

In the undated views seen, that above shows the engine in photographic grey against what is probably a cement backdrop provided especially when images of locomotives required to be taken. Certain components have been picked out in white and whilst the engine will be turned out in resplendent green for service, close examination reveals all the intricate aspects of the livery and lining are still present in this temporary livery. Details to note are vertical stanchions on the tender for the emergency communication bell / gong, the splashers and brake gear. Even at this early stage the standard train and brake whistles appear on the cab roof.

Opposite the photographer has turned his attention to the interior of the smokebox, the content of which is fairly obvious with the cylinder exhaust pipes combining to form the blastpipe and the blower ring and its steam supply at the top of this. Although the engine is externally pristine not so the interior of the smokebox and so it may well have been these were images taken either around the time of entering service or following an intermediate repair. *HL 01 031 (above) and HL 01A 107 (left)*.

6 TONS
CARR & WAGON ENGR'S DEPT
ACCUMULATOR VAN.
FOR THE CONVEYANCE OF
ACCUMULATORS ONLY.
PER PASSENGER
TRAIN ONLY
XP
DW 314 W

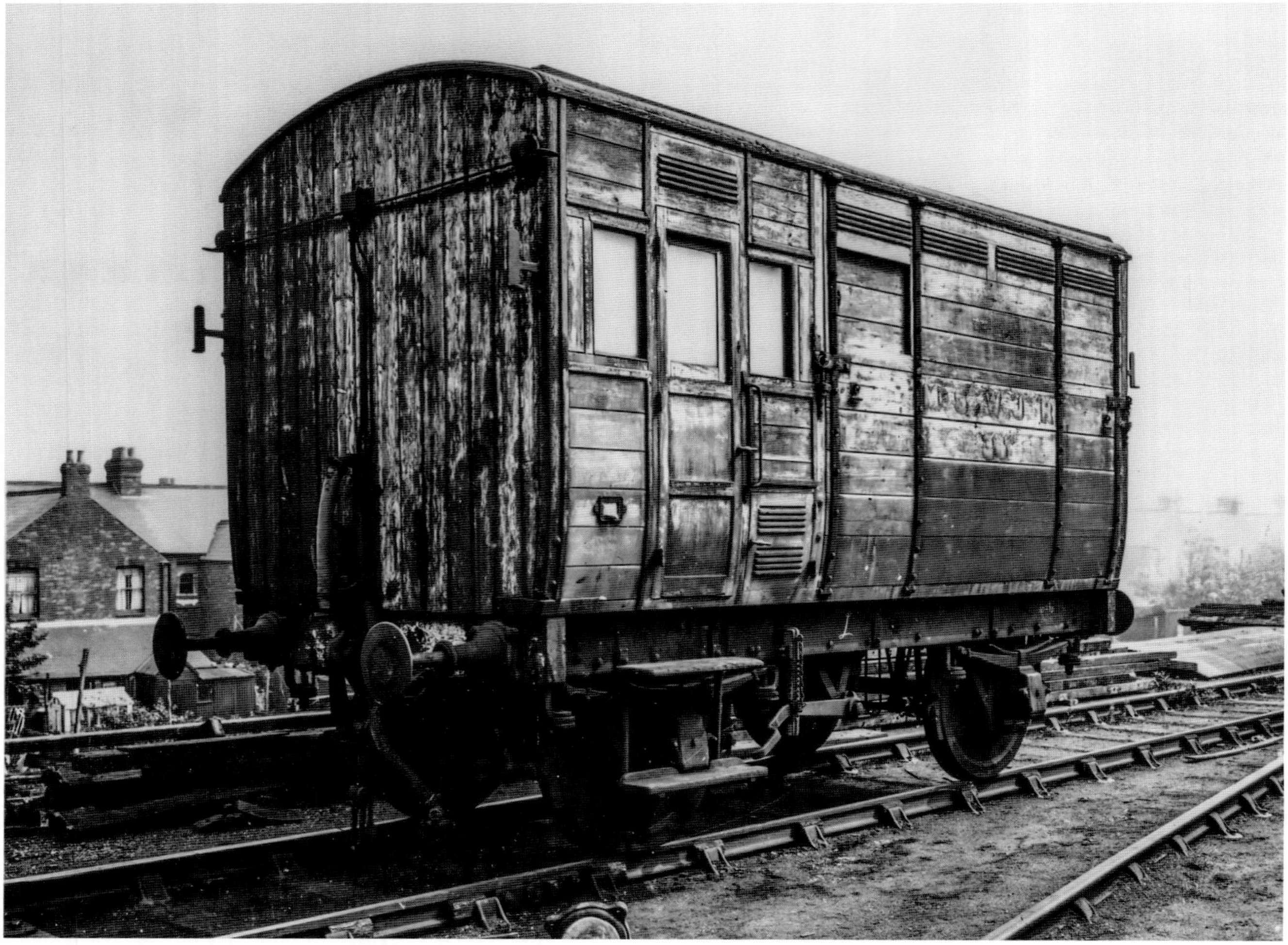

The unseen Great Western

Opposite top: 'Python A' No DW 314 W (obvious BR livery) and adapted for the carriage of accumulators. The origins of the vehicles in the Python series date from the late 19th century when the GWR built a range of covered and flat wagons for carrying private carriages; the days when by arrangement a party would turn up at the station and the train would include, usually at the rear, a van or open wagon on to which the carriage might be loaded. Sometimes a horse box ('Paco') might also be included in the formation. The first purpose built covered wagon for carrying motorcars, a 'Python', appeared in 1905 with the first 60 being 27' long and so able to accommodate two motors (a Rolls Royce or two...). Full width end doors allowed for the motor to be driven directly into the van via the station loading dock. They were classed as 'brown' vehicles with a livery that represented that colour and which because of their fitted brakes could also operate in passenger trains. Basically an early version of what BR would finally call their 'MotorRail' service. Some of the vehicles were fitted with dual braking meaning they could operate through to destinations where the Westinghouse brake was in use compared with the vacuum. Some Pythons were adapted for other purposes including circus trains with one vehicle, No 580 especially strengthened to carry elephants. In later years vans might be adapted for other purposes as here whilst in WW2 several had also been changed for use in Fire trains. The 'XP' branding was introduced on all four companies in 1938 and applied to 4-wheel vehicles having a wheelbase of between 10-15 feet. It signified the wagon was suitably braked and might be used in Class A passenger trains. *HC1 005 (48938)*

Opposite bottom: Midland & South Western Junction Railway horse box No 11. MSWJ expert Mike Barnsley gives the full details for this vehicle in his book on MSWJ rolling stock (Wild Swan) and where a full chapter is devoted to these vehicles. The No 11 seen was obtained second-hand from the Midland Railway in 1920 as a replacement for the original No 11. In total the MSWJ and its constituents had 44 horse boxes over the course of its existence 29 of which survived to be taken over by the GWR. Withdrawals started in 1924 and the last went in 1928. This, the second No 11, is seen here at Swindon in 1926 awaiting breaking having been withdrawn in December 1925 after renumbering as GWR No 27. It is just possible to make out the initials MSWJR on the planking mid-way on the side. *HC1 079 (53045)*

Above: Old Oak Common shed and the same area as the image involving Mr Cuss on page 6 - except here the grass definitely needs cutting. We may realistically comment that perhaps the depot is not quite open for business just yet, as witness the water column minus its bag, the various items of builders equipment and above all a distinct lack of locomotives. In the foreground the three open wagons are No 20520 a Loco Dept. 4-plank open, No 43089 a 10-ton steel coal wagon, and No 52923 a 4-plank open; again of 10 tons capacity. Those further away are not determined although it will be noted not all have either-side brakes. Dominating the scene of course is the massive water tower and coaling stage, the siding for coal wagons continuing on arches towards the main depot buildings. *HE2 017 (100986)*

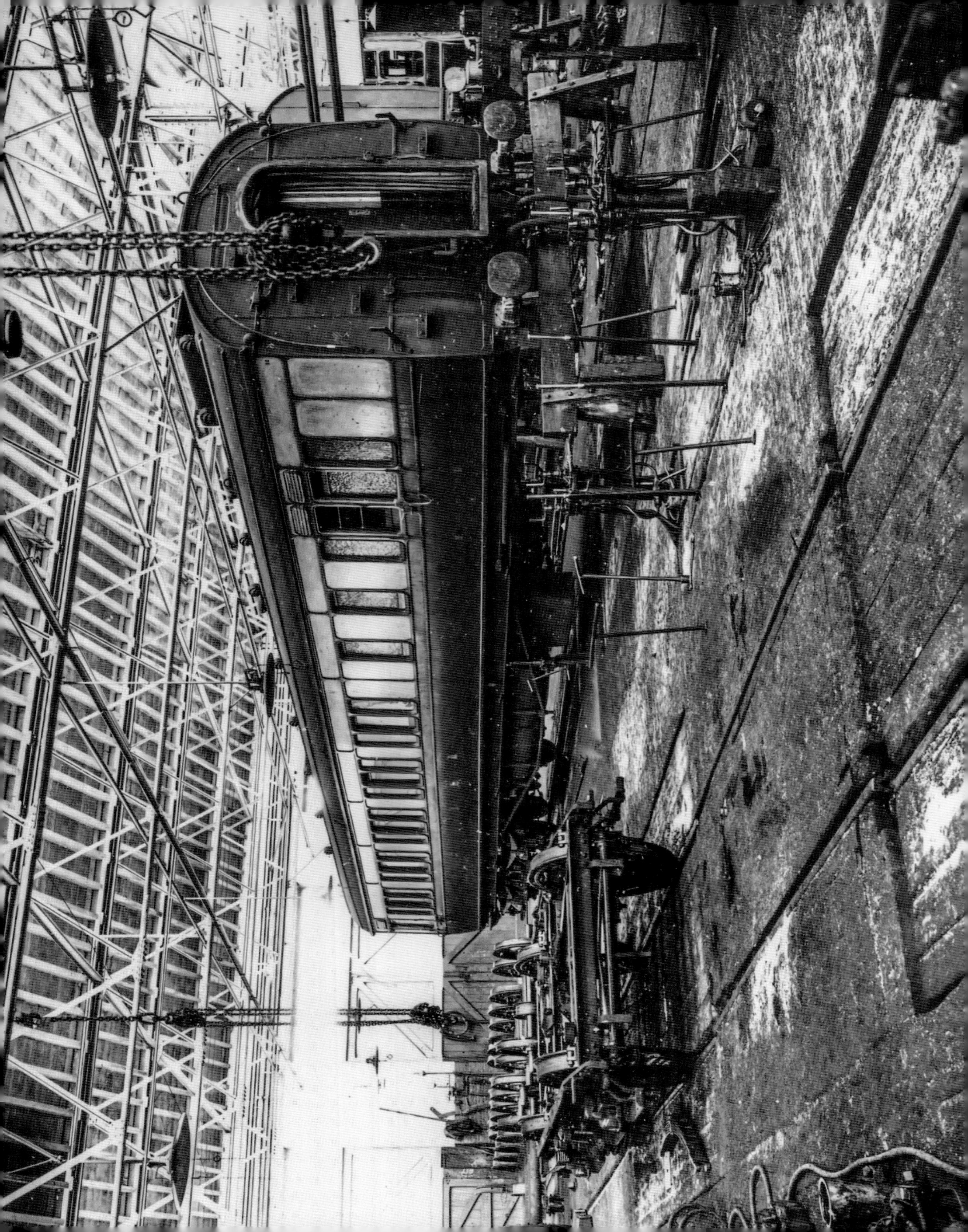

The unseen Great Western

Opposite: Views relating to the carriage shops are nowhere near as common as those in the locomotive works hence we were delighted when this was found in the archive. Even so it is not all it might seem as it was recorded in what was known as the 'Klondike' shop located on the east side of the Gloucester line. Here in 19C shop was where carriage bogies were changed, the original task being to change the bogies of convertible stock from broad to narrow gauge and naturally enough then known as the 'Changing Shop'. As to when the name change took place we may surmise as to being in the period 1896-99 and coinciding with the Canadian gold rush. (A description of bogie changing is contained in 'Special GWR Edition No 2 p129 - Wild Swan). In this image the vehicle shown is an example of Lot 1149 of 1 August 1908 a 57' Dining Car built to Diagram H16 and one of six such similar vehicles numbered in the series 9546 to 9551. This particular coach being No 9549. Post 1908 it would not be until the introduction of the main line articulated stock in 1925 that the GWR introduced any further non 70' dining vehicles. Diagram H16 coaches were in effect classless having two separate dining saloons, one seating 18 and the other 12. Seats were arranged either four to a table or two to a table, the GWR seeming to recognising that a little more elbow room was needed at meal times. No toilets were provided. Although built concurrent with the 'Toplight' style of carriages they were not Toplights in the literal sense of the word instead having filled in panels above the windows. Access was either by the corridors and adjacent vehicles or by a single set of doors at the end opposite to the kitchen; the kitchen located at one end as was practice at the time. As built they ran on 8' American bogies with cross springing, this according to Russell was changed to 9' bogies of conventional type supposedly in the late 1930s but may have been earlier. Possibly some updating of the interiors may also have been undertaken around the same time. Whilst the bogie change may not have been noticed by all, what did make a radical change was an alteration from the 'three windows' per table arrangement to a single window with a sliding vent, this took place in the 1950s. The photograph is dated 15 November 1927 and is described as a 'bogie lift'. If the bogie on the left is from the coach then standard bogies have either already been fitted or this is taking place. *HC7 008 (85700)*

Above: Another dining (Restaurant) car, this time No 9540, a 70' vehicle of Lot 1132 to Diagram H15 of which 12 were built by 31 August 1907 and carrying the running numbers 9534 to 9545. As with the H16 type opposite these started life on American bogies - which are fitted on the coach in the photograph - but changed to standard bogies prior to WW2. Again they were classless and could seat 42 in the four or two to a table style. Lighting was by electricity and cooking by gas; no doubt some succulent offerings were produced as well compared with today's microwaved food. They might be seen on any number of long distance trains where 70' vehicles were permitted. In the photograph above the roof board states, 'London (Paddington) Birmingham Shrewsbury & Birkenhead' whilst in Russell Part 2 (page 47) an illustration of the type carries the board 'London Newport & Cardiff'. Apart from a change of bogies, these vehicles also had their original three windows per table replaced with a single large window so giving a more modern appearance. A number of Restaurant / Dining cars saw out WW2 in sidings such as at Henley on Thames but were in a poor state by the end of hostilities. They were however restored to use and some at least were in main line use as late as 1961. One sad story relative to a GWR/WR restaurant car use concerns an Up train to Paddington where a passenger entered the dining car and asked the steward for tea and tea cake. He enquired if he might spend the remainder of the journey in the vehicle; this was agreed although the steward did add he would have to mind the staff working around him as they prepared the car for its return journey with the tables set for dinner. The train duly preceded on its way and when the passenger did not move upon arrival at Paddington the steward went to investigate. Poor soul had died en-route but had remained upright in his seat. As the steward commented later '...he did not even pay for this tea.' The GWR had several different types of 70' vehicles including, Toplight, Dreadnought and Concertina stock. In more recent times the latest IET sets have even longer intermediate vehicles which are 85' in length. *HC1 049 (53016)*

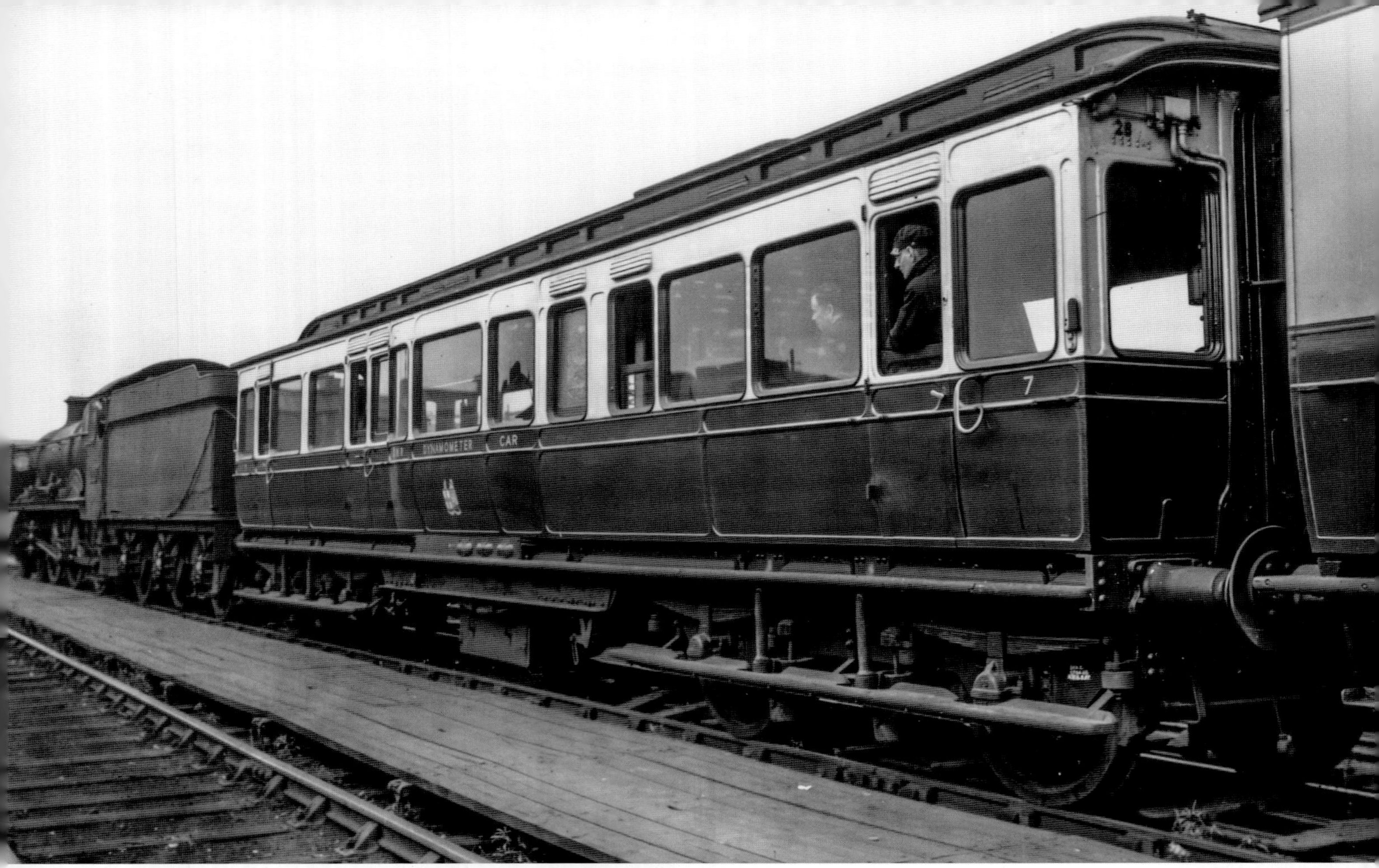
DYNAMOMETER CAR
7

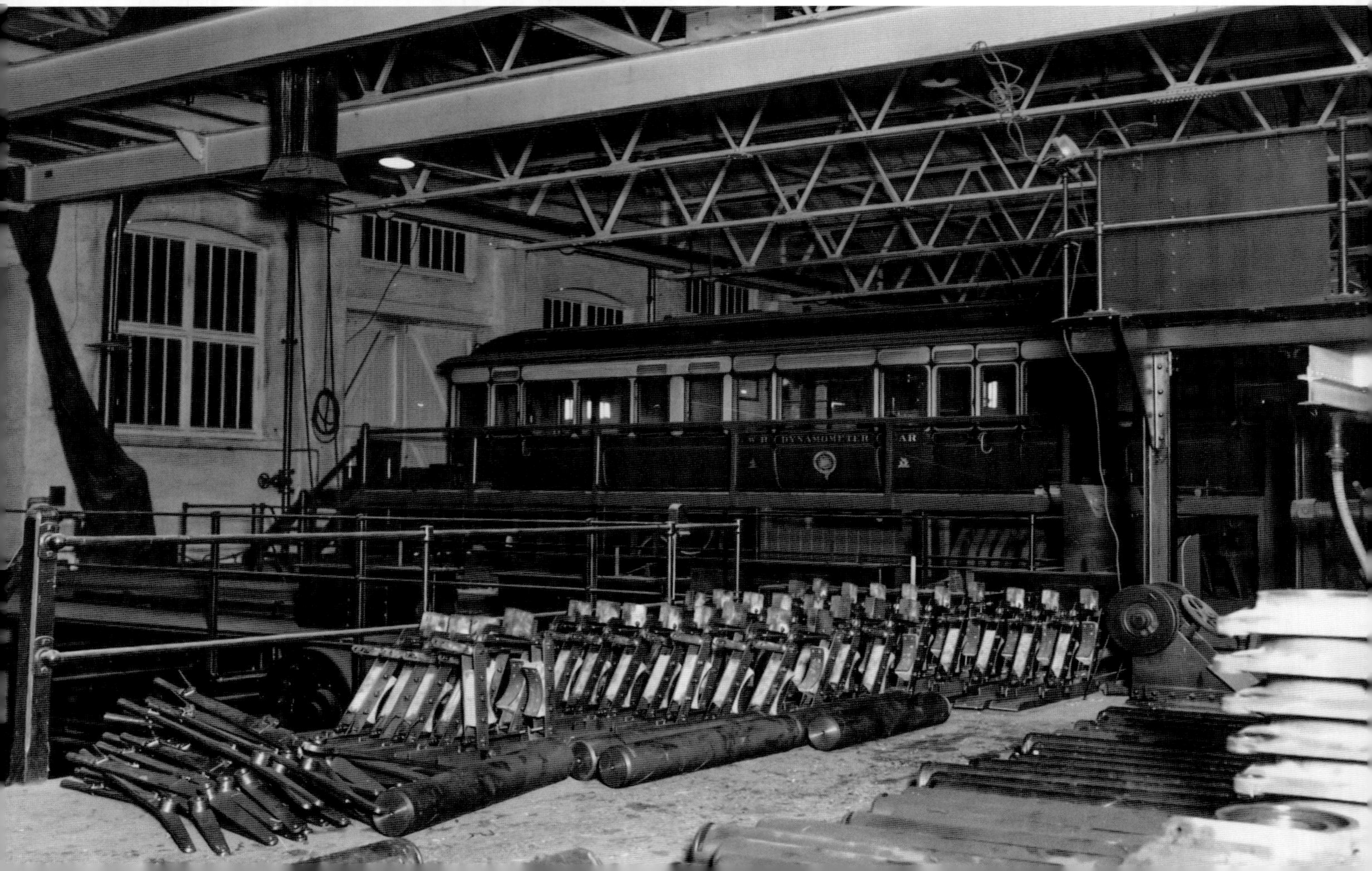
DYNAMOMETER CAR

Opposite top and bottom: The unmistakeable outline of coach No 7, originally GWR No 790 and finally in BR days W7. This was the unique dynamometer car built for Mr Churchward in 1901 to Lot 293 (which was an allotted wagon lot number), no diagram number was issued. Carried on two eight-wheel bogies the vehicle was 45' 0¾" long and 8 6¾" wide and cost £860 excluding internal equipment. So far as the bodywork was concerned, there was corridor connection at one end which was certainly present by 1924. The vehicle also had side look-outs and a 'Royal Clerestory' meaning the clerestory roof sloped at the ends. Externally there was also a retractable flange-less wheel accurate to 440 revolutions per mile. Its purpose was on-road locomotive testing to test efficiency in steaming, drawbar pull, indicated horsepower etc, for which purpose various measuring gauges would be attached to the trials locomotive. In this view some example connections may be seen in the cables from the car running along the side of the tender. The car saw use on the GWR behind examples of most of the major tender classes with these on the road trials additional to tests that could take place on the stationary plant within the works. (The stationary plant was upgraded in the 1930s to absorb the greater power as bigger engines were produced.) An example of use with the Dynamometer Car took place when steaming tests were carried out on No 4074 *Caldicot Castle* starting from Millbay sidings at Plymouth. In addition to the dynamometer car the engine was also fitted with a temporary wooden shelter across the front buffer beam where further observers could be carried, again recording various aspects. The top image shows a test involving a modified 'Hall' class locomotive and again with an indicator shelter across the front framing. Car No 7 also saw much use during the locomotive Interchange Trials of 1948 and later would run behind the Southern Region main line diesel No 10203 as well No 35020 *Bibby Line* during trials following rebuilding and during tests with 'County' and redraughted 'King' class engines. When not in use the vehicle was often stored within the works. Livery for most of its life was standard 'chocolate and cream' but there was a brief flirtation into 'crimson and cream' in the mid 1950s (before common sense prevailed!). Car W7 was eventually replaced by a purpose built conversion from a Hawksworth coach in the early 1960s (all third Diagram C82 No. 792 built August 1946, withdrawn c. 1961 and converted to Dynamometer Car No. DW150192; now preserved) with much of the internal equipment stripped from No 7 installed in its modern replacement. No 7 was eventually preserved on the South Devon Railway. The 'Royal Clerestory' style of roof was used on the GWR Royal train of the 1890s and also on some other specialist vehicles including the GWR directors saloon, latterly the Plymouth Divisional Engineers inspection coach No 80978. *Top: HC1 058 (53024) Bottom: HC4 005 (59034)*

Above: A slightly unusual process taking place outside the works, labelled as 'Testing of Carriage Warming Apparatus'. A steam supply is being taken from the locomotive on the right, a Pannier tank of some sort, which is then passed through various pipes to a sample carriage heater. The gauges will record the steam pressure, usually a maximum of 80psi and the pressure drop over distance; the reason why the front of the train was always slightly warmer than the rear. The purpose of the corrugated hose from the ground is not determined. It is not thought this was done with every new or repaired engine so may well have been a test of equipment. A partial date is given 26 August 195?. *(HC2 040 53123)*

Paddington Goods, undated but we believe early to mid 1920s and certainly prior to 1926 - see next page. Dominant is the Paddington high level coal yard, access to this was only by the wagon turntables seen, the railway crossing over the roadway with access to ground level by wagon hoists. There was limited interchange between the railway and canal by means of the 10T derrick crane seen in the far distance; Tony Atkins notes this was by 'prior arrangement'. Shunting horses were used to move the wagons around the high level yard. The name of one merchant is visible, 'T S & C Parry', with a number of PO wagons of the same name in the adjacent siding. The end of a Parry wagon also bears the letters 'W R D & S', as does the vehicle on the left. A vehicle on the right also carries a star on the end - Charles Roberts builders perhaps? Messrs Parry was just one example of a coal merchant operating in London at the time, a time when coal was the principal source of heating for most families and workshops and which would culminate in the great smog of 1952. Much redevelopment has taken place in the area in the subsequent 100 years, residential dwellings replaced or rebuilding following bombing in WW2. According to the official 'Station Diagram' track plans of the GWR, this high level yard could accommodate no less than 85 4-wheel wagons. *HG1 027 (81662)*

Companion to the view opposite is this scene taken immediately after a return to work following the 1926 General Strike. The strike had begun on 4 May and officially ended on 12 May but we cannot say this was the scene straight after as in many places there was a gradual return to work. Whatever, the image indicates a number of areas of interest; first the high level sidings have gone together with the connecting bridge, some building work is also taking place at ground level and which will culminate in the new goods depot in 1931; the former high level siding area becoming a vehicle park. We should also comment on the vehicles being used to move goods, almost exclusively horse drawn drays, although there is evidence of a small influx of motors on the extreme left. Sorting and movement of parcels and goods traffic was a highly organised but heavily labour intensive operation. What we now expect in our daily parcel deliveries from the various couriers that operate, was all done by hand and an army of checkers, clerks, loaders and labourers. The horses too had to be looked after and to this end the GWR maintained the Mint stables, the large brick building running right to left. Here worked labourers, farriers and a vet. As the use of horses declined so stalls in the stables were used for the storage paperwork from the offices at Paddington. The car man on the dray would also have a 'van lad' whilst other casual labour would be recruited on a daily basis from men who would stand outside the gates hoping to be picked for work. *HG1 024 (81659)*

The unseen Great Western

Above: Muscle power, the caption simply states 'GWR shunting horse pulling 10T loco wagon'. Exactly what was being demonstrated is not clear; the ability of the horse or equally its inability. Certainly the animal is straining into the task. The man by the wagon with his hand on the brake was essential as once the initial inertia had been achieved momentum would take effect aided perhaps by the gradient. At some rural locations the GWR continued with its use of horses beyond WW2 mainly at rural locations and with loads far in excess of 10 tons. Weather conditions could also make for a considerable difference in pulling power, a cold winters day when the axlebox grease or oil was sluggish not helping as would a frost. Local circumstances might also see the local coal merchant's horse 'borrowed' to place a wagon before going out on the round. *HG1 004 (72033)*

Opposite top: The expanse of Temple Meads good shed at Bristol. The advertising is akin to today's promise of overnight deliveries, a remarkable achievement back then. Although this is the GWR depot there are a number of 'common user' LMS vehicles. As one of the principal goods locations of the GWR it is not surprising to report there were seven GWR receiving offices in the City for goods and parcels. Remodelling in the form seen was authorised in 1923 and completed in 1929. There were now eight covered platforms and 14 sidings in the new shed which could accommodate 402 wagons under cover and 340 in the open yard. This compared with 242 and 200 for the old facilities. On the left and right sides were the cartage roads for loading and unloading. *HG1 035 (81670)*

Opposite bottom: Paddington parcels platform, basically an extension of Platform 1 and where parcels might be dealt with without the inherent disruption around the 'Lawn' as had previously been the case. This was another change brought about in the early 1930s, the new facility on the site of the previous extension to platform 1 - known then as platform 1A and used as the former milk platform (this dated from 1908). Vehicle access to the parcels platform was from Bishops Road, a ramp descending to platform level, the platform facing on to a separate set of rails parallel to those of Platform 1 as the latter left the station. The exit for road traffic was on to Orsett Terrace. Parcels were congregated in specific areas, those for 'Northern ' and 'South Wales' seen in the image. We have no date for the view but a study of the vehicles will indicate it to be pre-nationalisation and most likely mid to late 1930s - this also from the colour light signals at Paddington seen in the background which were installed in 1933. As with Bristol, there were numerous receiving depots in and around the London area, whilst road motors from the different companies would spend their day driving between each with specific loads. It was at this platform that the GWR diesel parcel cars might also be seen several times during the day. *HG1 049 (86128)*

The unseen Great Western

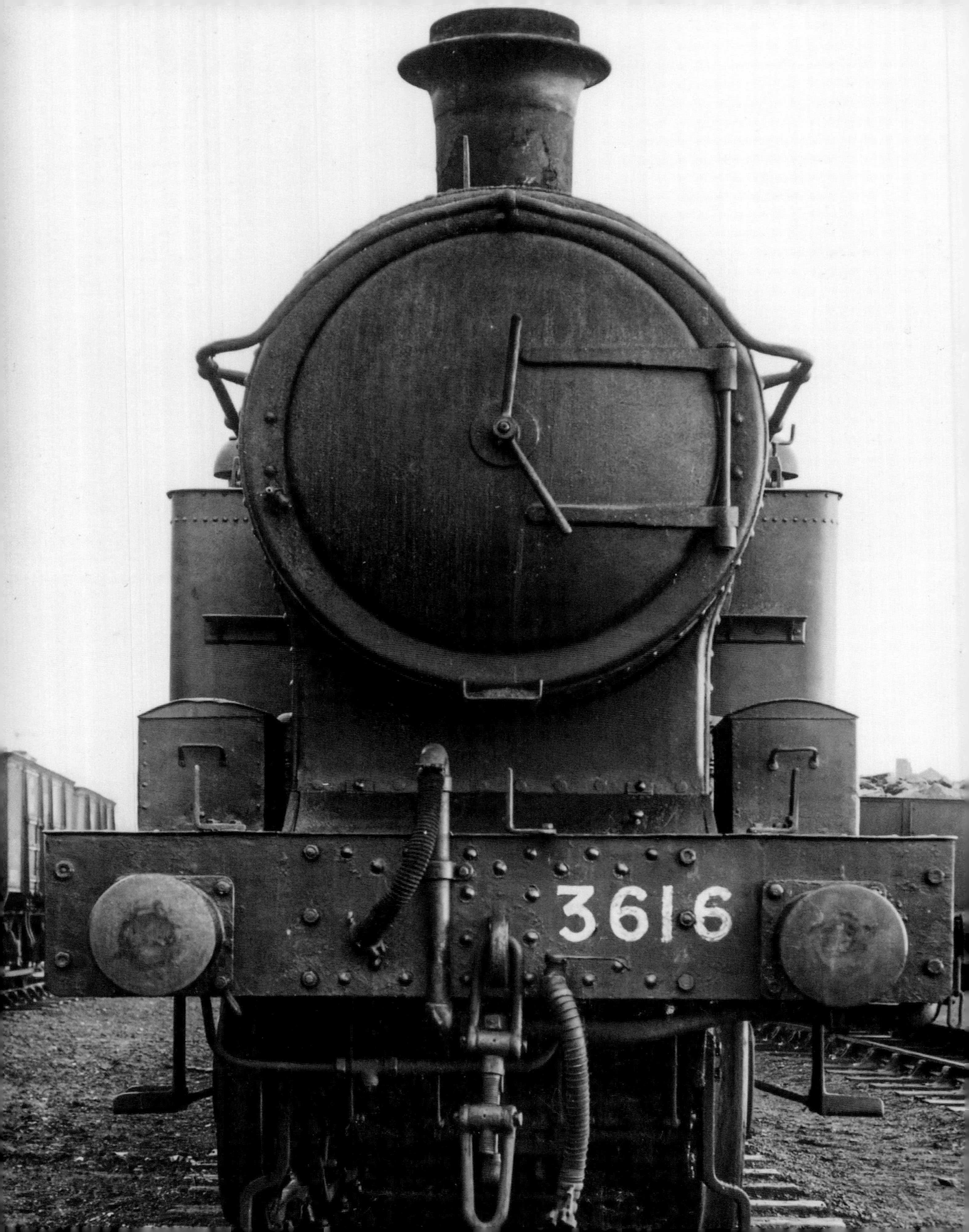
3616

The unseen Great Western

Opposite: Head on view of 36xx 2-4-2T No 3616. The origins of this design lay with William Dean who built an experimental 2-4-2T, No 11 in December 1900. This was an enlargement of the existing 2-4-0T 'Metro' (Metropolitan) design and proved immediately successful. In consequence an order for a further batch of the 2-4-0T design was cancelled and 20 engines to the 2-4-2T design were built in 1902. They took the numbers 3601 - 3620, the original No 11 renumbered as No 3600 in December 1912. Common between the prototype and the first (and second) batches built was a driving wheel diameter of 5' 2" and two 17" x 24" inside cylinders and a steam reverser. There was a slight variation in overall length between the prototype and subsequent engines, whilst the third batch, Nos 3621 to 3630 built in 1903 were fitted with taper boilers. Intended for suburban work in the London and Birmingham areas the class had commodious cabs with excellent vision for both forward and reverse running, resulting in their nickname of 'Birdcage'. (Damage to the windows at the rear from coal mainly when filling the bunker resulted in slightly smaller windows being fitted and coal bars added.) Slightly unusual for a tank engine was the fitting of scoops to allow for water to be picked up from troughs at speed. This caused an issue with No 11 on Rowington troughs; the inrush of water into the tanks was so great that the existing air could not escape quickly enough and the tank sides split. The remedy was to fit larger air vents to the sides and bunker. The class dominated the suburban scene until replaced by larger 2-6-2T types. For a while they might be seen in South Wales and also trains north of Chester. All were withdrawn in the early 1930s, No 3616 taken out of service in September 1933 and broken up at Swindon in January 1934. Its running number was bequeathed to a new pannier tank in March 1939. In the photograph there is a clear view of the toolboxes and also the enlarged vents atop the side tanks. *HL03C 033 (56811)*

Above: This is an interesting image for a number of reasons. Although appearing to be a 'Standard' or 'Armstrong' goods it is in fact a former Oxford, Worcester & Wolverhampton loco which in turn passed to the West Midland Railway and finally became GWR No 49 (formerly No 264). Note the substantial tie-bars at the back of the frames below the axle boxes. The tender is probably of the Dean 3,000-gallon type of which 397 were built between 1884-1906 or less likely a 2,500-gallon tender of which 301 were built in the years 1884-1903. Replacement of coal rails with side fenders commenced from 1903 but here it is in as-built condition. The tender is much younger than the engine as could easily have happened with tender rotation. The locomotive carries a round-top 'S2' type boiler. We do at least know the date of the image, 17 February 1911, the train an Up goods at Hockley. Notice the dumb buffer wagon immediately behind the tender; dumb buffer vehicles were barred under agreement with the RCH from main line running as from 31 December 1913. Even so there were still a considerable number around after that date so there may have been exceptions. The loco coal is also worth a second glance, some massive lumps around the side of the tenders forming a genuine coal wall - breaking one of those would not be easy. The signals seen are all 5' arms so refer to main lines and would be carried on wooden posts. It is quite likely the train was stationary when the view was taken; no driver at the spectacle plate or leaning outside the cab. The engine is carrying three headlamps across the bottom of the framing. This was the old code for a 'through goods, mineral or ballast train'. Lamp codes were changed from 1 January 1918 to represent what we were later more familiar with, the new code for this working removing the need for a lamp over the right hand buffer. *HL04B 013 (49633)*

We chose this view for the wonderful array of advertising shown on the exterior of what was for most of its life known as the South Devon carriage shed at Exeter St Davids. The railway companies were quick to identify the potential they offered to advertisers and in return the businesses concerned became aware how their product name or slogan might be seen by a greater number of members of the public. What we have in this image are examples of the colourful enamel adverts that once graced stations up and down the land, a few examples still to be seen on the various heritage lines often advertising products either national or local whose name or manufacturer has long faded into obscurity, such as Halls Distemper (a whitewash). Some names however continue into the 21st century under their original branding including Bovril, Whitbread, Pears soap, Wrights coal tar soap, Black & White whisky, Camp coffee, Emu brand Australian wine, Nestle's milk, the Observer, and Jeyes fluid. Others have either faded into obscurity or been merged with and into other concerns, examples being the Daily Sketch, and McKechnies whisky. There were two carriage sheds at St Davids, both located immediately south east of the station and parallel to the main running lines. Access was via a single siding that ran parallel with the steep connection between St Davids and the Southern's Queen Street station. The image we have is of the two road shed; the top of the three road shed may just be seen on the roof-line. In the view a part of a 4/6 wheel passenger vehicle may also be seen on the extreme right. This building appears on early plans as the South Devon Railway Loco Shed, presumably it originally housed the atmospheric/Moretonhampstead train sheds so may date from that period; respectively 1872 and 1886. Conjecture is also that it may also be the only three road train shed of that design. The access lines and sidings in this enclave were known as South Devon sidings right up until they were lifted. By the early 1970s the sidings and sheds were redundant and the latter sadly demolished. After this the site was initially used by Royal Mail, however once carriage of mail by rail ceased, the Royal Mail building was replaced by a veterinary surgery. Today the whole South Devon sidings area has been redeveloped as student accommodation, some named to commemorate the railway history of the location. *HC6 035 (85147)*

Loading sugar beet, it is said at Oxford in the early days of BR. Root crops such as beet were the traditional alternative to pasture, large quantities sent for processing; two plants being in GWR territory, Allscott near Wellington and Kidderminster. Sugar beet had been cultivated since the mid 1920s following shortages of imported sugar that had occurred in WW1. For many years there was a national sugar beet 'campaign' lasting for about three months in early winter with railway vehicles collecting the crop either from farm to the nearest railway station for onward transport by rail or if more convenient to Allscot and Kidderminster. Individual wagon loads could even be made up into complete trains, the load dense and in consequence heavy. It was not unknown for general purpose open wagons, even coal trucks on occasions, to be used. The vehicle seen here is a 10T open and it will be noted the brakes have been applied. Next is a 'conflat' probably previously loaded with a container. *HG1 019 (72048)*

The unseen Great Western

The unseen Great Western

Above: When referring to the South Wales ports it might be tempting to think of them dealing almost exclusively with the export of coal but that monopoly, if indeed it ever existed was destined to be short lived and the word 'diversification' instead came quickly about. Harking back briefly, the Queen Alexandra Dock at Cardiff had opened in 1908, its position parallel to the existing Roath Dock and on what was reclaimed foreshore. The water depth was a mean 37' deep, with 850' x 90' of dock area equipped with coaling cranes and warehousing on the north side. Steam coal had ceased to be the principal export commodity dealt with after the Admiralty moved to oil for navy ships following WW1 whilst the export market for coal to France similarly ceased after the treaty of Versailles. (This saw Germany paying war reparations to France in the form of their own coal.) Whilst the export of coal might have declined the facilities at the Queen Alexandra Dock grew to a total floor area of 378,000 sq ft based on six transit sheds. There was also a 33,000sq ft cold store. For some time Cardiff was also the principal British port for importing potatoes (the 9.45pm Cardiff-Saltney 'C' headcode express goods train was nicknamed 'The Spud'). The Cardiff wholesale fruit and vegetable merchant Messrs J E England wrote to the GWR in appreciation of the speedy discharge of just over 213 tons of new French potatoes from ship to rail in less than 5½ hours, the potatoes reaching destinations 200 miles away in less than 12 hours after the steamer had tied up at the Queen Alexandra Dock. Perhaps the most famous Cardiff export (and then import) in 1927 was No 6000 *King George V* the loco sent in August from the Roath Dock across the Atlantic to Baltimore docks for the centenary celebrations of the B&O on the *SS Chicago City*. The loco returned in October that year. Other traffic noted dealt with at the Queen Alexandra Dock included fruit from Spain in June 1933 and in November 1949 casks of asphalt. In the 1920s Cardiff attempted to rival Plymouth, Southampton and Liverpool as a sailing / destination port for the trans-Atlantic liners trade and in consequence the occasional passenger trains might also venture on to the dockside. In the view seen, sacks of barley are being unloaded ready for onward transit in ventilated vans. *HD1 076 (50192)*

Opposite top: This a view of the docks at Port Talbot whose origins may be traced back to 1834 although at that time under the name of the Aberavon Harbour Co. After initial business success, traffic slowed but the opening of the first part of the Rhondda & Swansea Bay Railway (between Pontrhydyfen and Aberavon) in 1885 fed shipment coal into Port Talbot from collieries along the line. Trade through Port Talbot increased further after colliery proprietors in the Ogmore and Garw valleys formed the Port Talbot Railway and Dock Co in 1894 intending to use the docks to export their coal. Following various improvement works the GWR took over the Port Talbot Railway, but not the docks, in 1908. That would come with the grouping in 1923 with Mr E Lowther General manager of the former PTR promoted to the position of Chief Goods Manager of the GWR as a whole. (Mr Lowther, later became the GWR's Chief Docks Manager.) Port Talbot was equipped with belt conveyors where a loaded wagon was tipped of its contents, the coal falling into a hopper from where it was fed on to a 3' wide belt and thus into a waiting ship. At Port Talbot, as at many other places in the coalfield, finely broken coal and coal dust was made into 'patent fuel' briquettes by combining with tar. Messrs Crown Patent Fuel works, Atlas Coke & Patent Fuel works and British Briquettes Ltd had factories around the dock at various times. Packages of patent fuel intended for export were loaded on to British Briquettes Ltd own flat wagons and run from the factory to the quayside where the wagon and contents was craned down into the holds of ships, unloaded and returned back to the rails on the dockside. The image however, displays what is believed to be silos for holding iron ore. Belt conveyors were also installed at Fowey by the GWR in 1909, this for the bulk loading of china clay. The image shows the *SS Albatross* along the docks at Port Talbot along with one of the briquette making factories alongside. *HD2 045 (50261)*

The unseen Great Western

Bottom: The conveyance of milk was an important traffic for the GWR. For some years this was carried in farmer's churns brought to the station and forwarded either by passenger train or if in quantity in specially built 'Siphon' vans, these of wooden construction and having slatted sides to encourage air flow. Of the four main line railway companies as existed in 1923, the GWR carried by far the most milk, followed by the LMS, the Southern and finally the LNER. The distinction simply explained by the prevailing westerly winds which brought rain and therefore the best pasture for cattle to the west. In the same year the total volume of milk carried was 282 million gallons. The milk was processed at local creameries before being distributed. Matters developed with rail tanks taking processed milk in train loads to the centres of population. Initially 4-wheel tanks were used but these were later superseded by 6-wheel vehicles able to run at faster speeds; and it was said less likely to 'hunt' from side to side so risking the contents turning to butter en-route. Milk continued to be transported by rail into the 1970s after which road took over, again the convenience of not having to load and then transfer the load the principal issue; even if this did mean countless more lorries on the road. The GWR and SR both operated 'roll-on roll-off' lorry trailer and rail wagon transport but this was still not considered as convenient as road despite the train taking the equivalent of perhaps 30 lorry tanker loads in one go. The view is taken at Royal Oak. *HG1 022 (81657)*

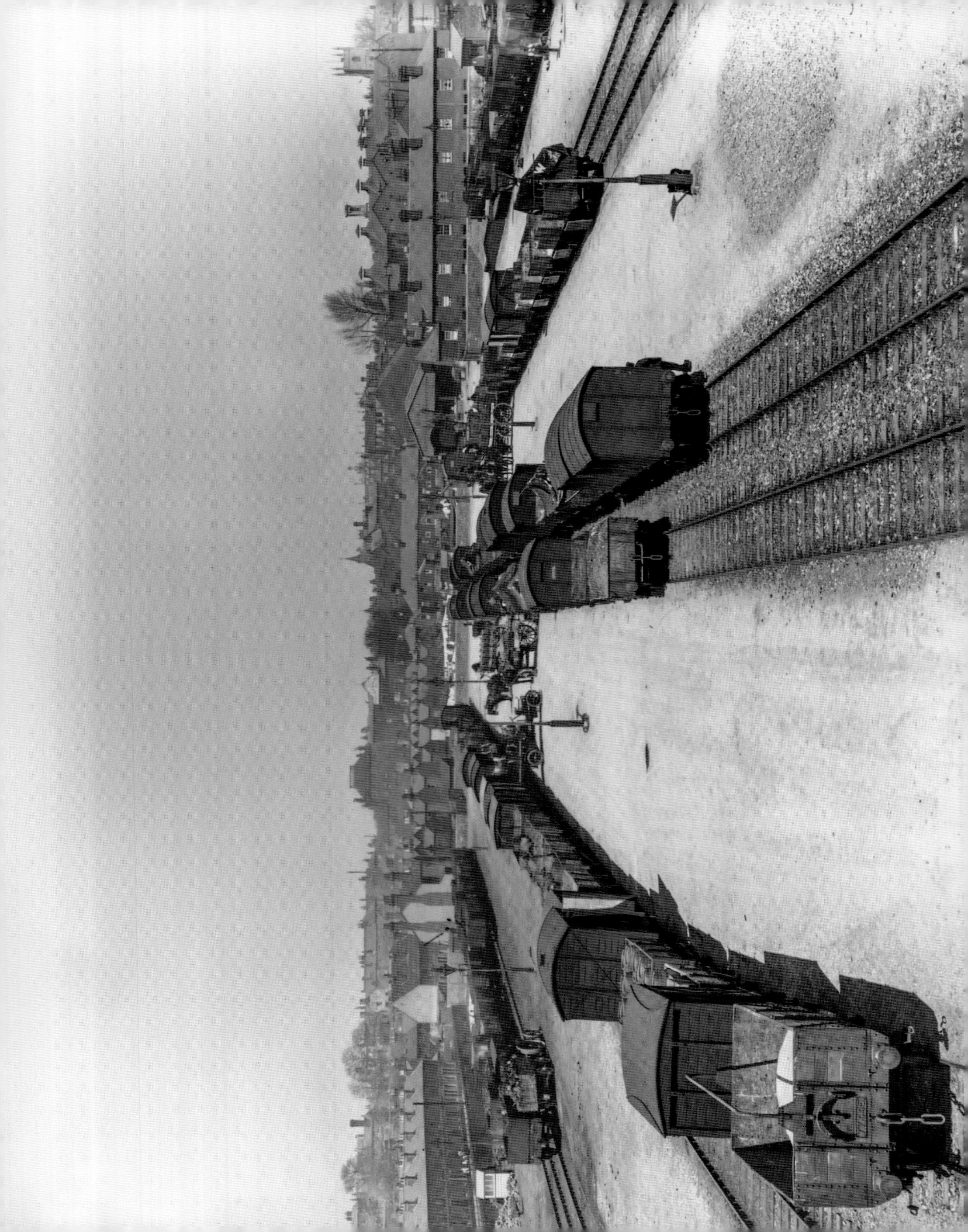

The unseen Great Western

Opposite: Reading Central goods yard at the end of the goods only Coley branch which diverged east off the Berks and Hants line at Southcote Junction. The Coley branch line was 1 mile 61 chains in length and was built by the contractor Henry Lovett of Wolverhampton on behalf of the GWR; being authorised by the Great Western Railway (Additional Powers) Act of 1905. It opened to traffic on 4 May 1908. Road access was from nearby Fobney Street. The purpose of the new yard was to reduce congestion by diverting local goods away from the main line and at the same time reduce cartage by providing an additional goods facility on the opposite side of the town centre to the existing Vastern Road yard; the latter north of the main passenger station. Apart from here at Coley and the Vastern Road yard already mentioned, there was an additional yard with the main goods shed. This was alongside Kings Meadow Road just east of the passenger station at a lower level to the main line. (Kings Meadow also dealt with quantities of biscuits from the nearby manufacturer Messrs Huntley & Palmers.) Coley was built partly on a site previously used as wharfage by the nearby Kennet & Avon canal, the GWR responsible for having to re-site a Masonic Temple from the location. Sorting of arriving freight trains and necessary re-marshalling was undertaken either at the West Junction Marshalling Sidings or the exchange sidings at nearly Scours Lane. The yard was busy from the outset, made even more so in 1910 when it was the railhead for unloading livestock and machinery for the 1910 Royal County Show. No passenger facilities were ever provided although in its final years it was visited by various enthusiast special workings. Within the yard were twelve sidings located in pairs with good hard-standing which could accommodate approximately 300 4-wheel wagons. In addition to the main yard there were additional sidings en-route serving a jam factory and a sawmill. From the main yard one siding originally continued north-east to H & G Simonds Brewery. At Bear Wharf, adjacent to County Lock on the canal, another siding was built to allow trans-shipment of goods between rail and barge, locomotives sometimes being used to tow vessels short distances against the current. One perhaps strange omission was there was never any provision for watering locomotives and certainly in later years and possibly throughout its existence, the twice-daily trip workings were always undertaken with tender engines which would also shunt the yard as necessary. (Apart from 'King' class engines classified as 'double red' the branch could accommodate any type of engine. The view dates from the period prior to WW1 with steam (traction engine) power, horse power, and road motors seen. When viewed under a glass, in the right background can be seen the covered shed 'GWR Central Goods Station'. In front of this was a 177' end-loading ramp / loading dock on which was mounted a 30cwt crane. Elsewhere was a 10T loading crane and a 20T cart weighbridge; the office for which is just visible on the extreme right. As this was a single line goods siding no signalling was provided except where it met the main line at Southcote. The points at the intermediate private sidings were controlled by Ground Frames released by a key on the branch staff kept at Southcote Junction signal box when not in use. Within the yard itself turnouts were simply hand operated. After MAS (Multiple Aspect Signalling controlled from a single location) was introduced in the Reading area, the final years were under Western Region C2 working (lines worked and controlled by a 'nominated person'). The mass increase in road transport saw the depot's use decline and all facilities were withdrawn in July 1983. *HG1 007 (72036):*

Above: The Kings Wharf Government cold store at Cardiff. This was one of 43 identical emergency cold stores built around the country in WW2 and in use by 1941. In addition there were an unreported number of 'buffer' (strategic reserves of food) depots, some possibly included in the above number. According to Subterranea Britannica, a further 40 grain silos were built around the same time. Most were sited away from primary target areas although obviously with the exception of the sites here at Cardiff, and also Wolverhampton and Aintree. All were located for ease of rail access often with new sidings provided. Food storage at the request of the Government continued well beyond WW2 and into the years of the cold war but ceased around 1961. In the view seen contemporary road lorries are seen along with a steam heated banana van. *HD1 027 (50165)*

The unseen Great Western

Above: The post-war GWR venture into oil burning has been touched on in other publications but now may be the time to afford some detail perhaps not previously recorded. Neither was this the first nor the last time GWR locomotives would be converted to burn oil as back in June 1902 No 101, an experimental 0-4-0ST had been built at Swindon specifically to burn oil fuel. It was converted to burn coal in 1905 and in this condition withdrawn in 1911 after a short life of just nine years. Post WW2 the GWR were experiencing difficulties in securing quality steam coal as had been used before; much of the latter being sent for export to gain essential foreign currency with the legacy that engines were failing out on the line through shortage of steam or excess firebox clinker including, in of all places South Wales. It was in consequence of this and also to resurrect their own pre-war ideals of abolishing coal haulage to depots in the West Country that in 1946 the railway embarked on an ambitious plan to convert engines of various types to burn oil with the ultimate aim of making Cornwall a coal free area. (The only known drawing of an oil-burning tank locomotive is of a Class 4575 so equipped. This class was a popular type on Cornish branches and this proposal would have fitted within the strategic intent regarding the West Country.) To test the scheme the depots at Swindon, Didcot, Reading, Westbury, Old Oak Common, Gloucester, Bristol Bath Road, Bristol St Philip's Marsh, Gloucester, Severn Tunnel Junction, Newport Ebbw Junction, Cardiff Canton, Llanelly, Newton Abbot, and Plymouth Laira were provided with oil fuel bunkering facilities; the work at some depots proceeding faster than at others and not all were fully operational before the scheme was subsequently abandoned. Work at two other locations, Banbury and Swansea was also cancelled before completion. Eighty plus years later almost the only physical reminders of the scheme that remain are the concave supports for oil tanks that exist at the Great Western Society shed at Didcot although the tanks themselves were removed long ago. Notwithstanding the comment over the aim of having Cornwall as a coal free area it was clearly the intention to fully test the scheme before progressing to Cornwall. A total of 37 steam engines of various classes were converted to burn oil, the most numerous of these the 12 members of the 28xx / 2884 type which were renumbered in a series starting 48xx. Eleven members of the 'Hall' class were similarly converted and renumbered in the 39xx series. (At that time the 48xx number series was already allocated to the 0-4-2T auto/branch engines, this whole batch hastily renumbered in the 14xx series but with no change to their fuel - coal.) Other engines converted will be referred to over the next few pages. The conversion of the 39 steam engines took place at Swindon between June 1946 and August 1947, the work progressing concurrent with the necessary depot facilities. In the view above the renumbered 2-8-0 No 4855, one of the final engines to be converted in July 1947 is seen 'ex-works' alongside the facilities at Swindon running shed. This engine had been built at Swindon as No 3813 in September 1939. It was restored to its original number (and coal burning) in June 1949 and survived in service until July 1965. *HL08B 010 (68620)*

Opposite: The re-fuelling point at Swindon running shed. One siding has been devoted to supply oil with two counter-balanced delivery arms and fixed ground level oil storage tanks beyond. The Anglo - Iranian Oil Company assisted Swindon in the conversions affording advice where necessary. The overhead delivery arms here are interesting as widely seen official publicity photographs show the tender tanks being replenished at rail level courtesy of a pipe screwed directly into a similar fitting at frame level. Possibly then tenders could be replenished in either fashion. At ambient temperature the oil used, basically 'Bunker C' was hardly viscous hence the ground storage tanks were steam heated along with suitable insulation - as seen- surrounding the ground level delivery pipes. The tanks fitted into the former coal space within the tenders also had a series of steam heating coils fitted again to maintain viscosity. Not mentioned but almost certain, was the facility to supply an external steam source to a stationary (engine and) tender should an oil fired locomotive be out of service for a period. Without this the oil simply would not flow and it would not be possible to light the burner in the firebox. At this time, all the GWR locomotives converted were of the tender type, plans to convert a number of 42xx 2-8-0T engines not proceeded with. *HE2 028 (59491)*

The unseen Great Western

The unseen Great Western

Referred to as a 'stand pipe' and interesting as there is no flexible delivery pipe at the end of the horizontal delivery tube. Heavy oil was an unpleasant, noxious substance, emitting a strong odour and quickly contaminating any surface it leaked on to. That said one of its principal advantages came in relation to its calorific value, approximatively 18,000 BTU's compared with 14,000 BTU's for the best quality steam coal - and engines were certainly not always supplied with the latter. In consequence an engine required less fuel compared with coal to perform the same task whilst steam production could also be increased by more than 20% if required. All this was with little effort from the fireman who would be at his controls adjusting the rate of oil flow and steam into the firebox which in turn atomised the oil so creating the conditions for ignition. If the oil supply was judged correctly there was just a slight brown emission at the chimney, if judged incorrectly a heavy clag would descend. *HE2 030 (71609)*

The unseen Great Western

Above: Here we have a posed view of the unloading of a tank car into the site storage tanks; the 'portable trolley' to hold the hose may be noted. The tank contents would cool during transport hence the need for a steam supply into coils within the tank - this is seen in the small hose connected. A pump house was also needed and which may well be the building close to the end of the tanks.
HE2 023 (59492)

Right: Storage tank, capacity not reported. As they started early, the GWR were far ahead when it came to the Government rolling out the scheme to the other railway companies. Indeed the Southern Railway were the only ones to really grasp the potential although none of their depot facilities matched those of the GWR. Unlike on the GWR, when the scheme came to an end the Southern did not revert all their converted engines back to coal leading to a shortage in motive power. The concrete tanks supports seen and mentioned previously, are the type that have survived at Didcot.
HE2 027 (69687)

39

The unseen Great Western

Taken on 28 March 1947 here we have Castle class loco No 5083 *Bath Abbey* being refuelled at Swindon. The engine is in its final GWR livery and was converted to burn oil in December 1946. Five Castle class engines were modified to burn oil, Nos 100 A1 *Lloyds* (January 1947), No 5039 *Rhuddlan Castle* (December 1946), No 5079 *Lysander* (January 1947), and No 5091 *Cleeve Abbey* (December 1946). Each was attached to a 4,000 gallon water capacity tender with the coal space now occupied by an oil tank containing 1,800 gallons of oil and shaped to fit into the existing coal space. Note the oil tank sides slope inwards at the top to conform to the loading gauge. In this view the crew are jointly filling the tank; no doubt the top also extremely slippery; access to the tank filler was presumably via the rear steps as would be for the water filler. (This tender does not appear to have a frame mounted filler point.) The first three oil-fired 4-6-0s entered service coupled to 3,500 gallon tenders; neither Nos. 5039 or 5091 for very long, while No. 3950 *Garth Hall* was so equipped well into 1948. The two Castles were the last known examples of the class to work with small tenders in ordinary service. We are not informed how much time it took to refill an empty tank. Note also the size of the storage tanks and what is steam issuing from further down the site. In addition to the 2-8-0, Castle and Hall classes conversions, one 2-6-0 No 6320 ran as an oil-burner between May 1947 and August 1949. Apart from the obvious tender modifications, the engine fireboxes were altered by replacing the fire-bars with a plate that had openings for the air supply, and lining part of the firebox with high alumina fire-bricks to cope with the potential for rapid changes of temperature should the burner not be lit for any reason. Following experience in service the burners fitted were of the Laidlaw-Drew type. Adding additional fire-bricks had the effect of reducing the firebox area (- we are not told by how much), but this was compensated for by the increase in efficiency when burning oil. In theory it was possible to shut off the oil completely on a down grade or when stationary and so operate without using any fuel. Re-ignition was either by restarting the atomised supply of oil into a hot firebox which should, in theory then automatically combust (having reached its flash point), or by inserting a burning rag through the inspection hole in the small space where the fire-hole doors had once been sited. (Hopefully no GWR driver and fireman were faced with the problem that occurred on the Southern Railway where a footplate crew having let the burner go out discovered neither had any matches on them to light a rag; an unscheduled stop was made at the next signal box.) In reality re-ignition did not always automatically occur as expected and flashbacks were commonplace especially where a residue of unburnt oil had built up. Fireman thus had to learn to judge their oil flow rate exactly as they would when using coal. Daily locomotive maintenance was also reduced as there was no need to clean the firebox, ash-pan and smokebox, although it was found necessary to clean the firebox atomizers daily. Tube cleaning was also required as oil residue could collect on the inside of the tubes reducing the conduction of heat to the boiler water. In service the oil fired engines were popular especially with the fireman whose manual labour was considerably reduced. Where they were not popular was when it came to dropping unburnt oil on to the sleepers and rails which could cause adhesion problems for a following train as well as enhancing the risk of fire should hot ash and cinders be dropped by another loco. Long term maintenance costs relating to additional firebox wear caused by extreme sudden changes of temperature would tend to indicate far higher maintenance costs for fireboxes and boilers on these engines. Despite contemporary publicity to the contrary, the mileages run were also low by comparison. A further issue arose from complaints by footplate crews that the engines were cold to work on with of course little residual heat from the firebox. *HE2 022 (59493)*

The unseen Great Western

Quarter ended	28xx	2884	Hall	Castle	63xx
Dec-45	2	2			
Mar-46	1	1			
Jun-46	4		1		
Dec-46				1	
Mar-47				2	1
Jun-47	1	1	11	2	
Sep-47	4	3			
Total	12	7	12	5	1

This is a most interesting view depicting No 4808 (renumbered from No 2834) in steam as an oil burner but without a tender. The engine is also clearly fresh from works so this may well be a test steaming of sorts. The data sheet within the Harry archive for this view gives a date for renumbering as seen in November 1946, whilst published records indicate this engine was converted in July 1947. Both may be correct / incorrect likely depending on when the engine was perhaps first taken into shops and which record was consulted. As an aside the tender in front belongs to No 77000, an Austerity 2-8-0. Note too in the cab the fireman's seat, previously a simple tip-up wooden affair, has been replaced with a box and padded seat. This is where the fireman would be located for most of the journey although he would still have to get up from time to time; for example to turn the water supply to the injector on/off, the control handle for which was situated on the tender. As mentioned, twenty 2-8-0 tender engines were converted to the new number series running from 4800 to 4811 for engines from the 28xx class and 4850 through to 4857 for those of the 2884 type. New numbers were allocated as the engines were converted and did not follow a consecutive order when compared with the original numbering. As mentioned, the GWR had started converting engines to burn oil, plus the refuelling depot installations in 1946 and with the first conversion, No 3950 *Garth Hall* (formerly No 5955) commencing work in June. We may be certain the government were aware of this progress as at the end of August the Minister of Transport announced he had authorised the main-line railway companies to proceed as quickly as possible with the conversion of 1,217 locomotives from coal to oil burning. The breakdown of this was GWR 172 engines (37), SR 110 engines (29), LMS 485 engines (17), LNER 450 engines (1). The figures in brackets show the actual number that were converted. The August 1946 statement commented work was to begin immediately. Thus from August matters progressed, the number of conversions indicating the degree of progress made although delays occurred as equipment was not always immediately available from outside sources. This especially applied to the depot installations. Certainly on the GWR and SR trains were being regularly operated by locomotives burning oil fuel, diagrams having to be especially prepared to allow for refuelling to take place. (It is possible that at some locations on the GWR refuelling plants were not fully ready to be used and as on the Southern Railway recourse may well have been to direct pumping from rail tank car to tender utilising an external pump and similar external steam supply. This was certainly the case on the SR at Eastleigh.) Unfortunately it was around this time that the Government announced - and despite gaining foreign currency by selling steam coal - they had insufficient funds available to continue the purchase of oil; the price of both exported coal and imported oil experiencing volatility in the markets. Accordingly a general order to all four railway companies was issued on September 23 1947 that further conversion work was to cease. The final GWR engine converted was No 4811 (2847), a 2-8-0 in September 1947. Four months later on 8 January 1948 another formal order was issued, this time to cease work on both converting engines and progressing the ground installations. Officially referred to as a 'postponement' work would never restart again. Up to that time monthly progress information on the rates of conversion and miles run by oil fired engines had been required by the Government. Now instead there was a request from Government for the costs to restore the converted engines back to coal burning. In the event the Government would reimburse the railway companies the cost of all locomotive conversions - excepting that is the GWR who would have no recompense on five engines converted before Government had officially approved the plan. (*Garth Hall* was certainly amongst the first five but details of the other four are not confirmed as 12 other locos are shown as converted in 1946 but all in November / December. Reimbursement of costs so far as the ground installations were concerned is not clear.) All the GWR engines converted reverted to coal firing in the period September 1948 through to April 1950 and where renumbering had taken place, a restoration to the original number occurred. In many respects it was an unfortunate case of the GWR leading the way and then being penalised for their own success. *HL08B 043 (61653)*

The unseen Great Western

Another most interesting image. This is 2-8-0 No 2872 but with an oil tank. The number of this engine would change to 4800 soon after and so on paper was the first of the 2-8-0 conversions. The date for this is reported as November 1946 but as described on the previous page, we cannot always take the conversions dates to be strictly accurate. The conversions also had additional connections between the engine and tender for the oil-burners (steam supply to the heating coils in the tender), and modifications to the firebox and of course the oil tank within the tender. Following the Government order to cease work on conversions, trains continued to be worked with converted engines until literally the oil ran out. This was by now in the time of the new British Railways and whilst dieselisation was on the horizon it was not considered that the existing depot installations could be adapted to suit diesel traction. Accordingly the diesel railcars and diesel shunters operated by the (G)WR continued to refuel as before. Heavy oil fuel would be used once more on what was now the Western Region. This was on the initial gas-turbine No 18000 which continued with this fuel throughout the whole of its life. (This excludes the small diesel fuel engine fitted to No 18000 and utilised when starting.) Of passing interest is when in April 1958 57xx 0-6-0PT No 3711 was converted to burn oil, a rectangular bunker fitted into the former coal space. The work was done by Messrs Robert Stephenson & Hawthorns. No 2872 was originally built in November 1918 and survived in traffic until August 1963. *HL 082E (98703)*

The unseen Great Western

Cab view of No 5083 *Bath Abbey*, **converted to burn oil in December 1946 and restored to coal November 1948.** The view shows the 'fireman' sat on his padded seat (no such addition was provided for the driver who still had to make do with a tip up wooden seat). The firedoors have effectively been sealed although at the top there is a top inspection hole available. The fireman's controls deal with the oil and steam flow. Beneath the footplate the insulated pipe carries the oil to the firebox burner. (The fall plate propped on to pieces of wood for the benefit of the photograph.) The view is clearly posed 'cold' as the steam gauge - top left - reads 'zero'. Incidentally do note the regulator handle and reverser are not painted. Prior to nationalisation it was simple burnished steel. The trend to paint such items in red only came in post 1948. *HL13D 065 66093*

43

The unseen Great Western

Above: The shed for Fishguard although often referred to as being sited at Goodwick as the latter was the nearest station. Opened in 1906 in conjunction with the new line and harbour at Fishguard, this was the servicing facility for engines having arrived at the terminus as well as having its own allocation. In this view the depot is seen probably within a few years of being opened, the motive power representative of the engine types (not necessarily classes) that might be seen over the years. Visible are a curved frame 'Saint' 4-6-0 and a '517', the latter No 1161. On the turntable could well be a 4-4-0 and there is another tender engine within the shed itself. Of note is the neat and tidy appearance; ash piled in wagon No 9676 whilst a number of full 20T coal wagons are lined up on the coal stage road. *HE1 093 (81643)*

Opposite top: There were two engine sheds at Neath, one the former Neath & Brecon railway depot and this larger shed referred to as 'Neath (Court Sart)'. This was another depot having two roundhouses each with 24 roads. Adjacent to the depot was a carriage and wagon repair shop. A new coaling stage and ramp surmounted by a 90,000 gallon water tank was added in 1921 and it is likely this view was taken around that time. Engines might be coaled on either side of the facility. The shed offices are visible on the extreme right and there are additional offices and stores in the building directly in front of the shed beyond the coaling stage. The allocation was one suitable to freight working and consisted mainly a number of 2-8-0 and 0-6-2T locos plus of course pannier tanks. In the view above we have GWR and ROD type 2-8-0's plus pannier tanks and 517s class 0-4-2T engines from a bygone age. *HE1 016 (64831)*

Opposite bottom: There were three known engine sheds at Taunton. The first opened in 1842 by the Bristol & Exeter company. At Taunton as indeed elsewhere, engineers were learning through experience, the first station here a 'one-sided' affair with wagon turntables and crossovers separating the Up and Down platforms and buildings. A two-road covered engine shed was provided just south east of the station with associated buildings including a Smithy and store. A small turntable was provided but from plans (*GWR Engine Sheds 1837 - 1947* by Mountford and Lyons) this would appear to show that a tender engine would have required to have been separated from its tender with both turned separately. This shed closed in 1860 and was replaced by that seen. By 1885 the station had also been rebuilt with an overall roof and directional platforms opposite each other in conventional style. Again the replacement was a two-road affair with offices partly occupying one side and a turntable on the station side of the shed with 'coal plant' nearby. We may take the term 'coal plant' to mean a coaling facility and most certainly involving human muscle power. The white building to the right is not identified whilst it will be noted the track in the vicinity, all sidings, retains the baulk road. The view seen is purported to date from 1899 but this may be questioned as a third locomotive depot in the form of a roundhouse was opened in 1896, opposite the south east end of the station. It is possible the 1899 may then be incorrect, alternatively this was a record view taken prior to demolition. Alternatively the old engine shed may well have found a further use for a time after the new roundhouse came into use. The third shed at Taunton closed in October 1964. *HE 102 (81649)*

The unseen Great Western

The unseen Great Western

The unseen Great Western

Opposite: Star class 4-6-0 No 4046 *Princess Mary* having its front framing blasted at Bristol Bath Road depot. Built at Swindon in May 1914, this engine had a 37-year life until withdrawn from Shrewsbury on 15 November 1951. The view is undated but we are certain taken before January 1949 as this was when external elbow steam pipes were fitted. The engine is standing over a pit road; most sheds had hosing down equipment fitted in the area of the ash roads in or before WW2 so that red hot ash might be rapidly cooled to prevent attention from enemy aircraft. The GWR along with at least the Southern experimented with both fixed and portable cleaning equipment, pre WW2 in an attempt to improve efficiency and post WW2 due to labour shortages. *HL11B 031 (68399)*

Above: What we would now refer to as power washing taking place on a 28xx at Old Oak Common on 23 July 1930. Alongside, the procedure is being observed from the footplate of No 4041 *Prince of Wales* (built at Swindon in June 1913 and withdrawn on 19 April 1951.) Careful observation reveals two hoses connected to a ground source and with an insulator wrapped around what is the steam pipe. The steam would be sourced from the stationary boilers at the depot. Adding steam and water in varying quantities could blast away oil and grime and whilst it would never equal the kind of care lavished on engines in the Edwardian era it did at least remove much of the muck. A second artisan is undertaking the same process on to the inside of the frames from footplate level. Both are also well attired in waterproofs, no doubt there were several blowbacks. Whilst cleanliness might be taken as only having an aesthetic advantage there was also a more serious side as general work-a-day much could accumulate on components so hiding the potential for a future failure. This may then have been a simple cleaning exercise or one where the intention was to identify a reported issue. The engine will also be noted to be in steam and will have to have its inside motion and other oiling points thoroughly lubricated again before its re-enters service. Whether this type of cleaning was commonplace, or limited to Old Oak and indeed how long it continued, is sadly now lost in the mists of time. *E1 086 (71995)*

Copies of the images within the book are only available from **STEAM** Swindon.

Please quote the subject, page number and reference shown at the end of the relevant caption.

The unseen Great Western

The unseen Great Western

Opposite page, top: GWR 'Dean Goods' No 2559 photographed after an altercation of sorts, details of which we are not given. No 2559 was one of 30 engines of this type built to Lot No 111 in 1897-99 and were the final engines in this 260 strong class. No 2559 displays accident damage at both the rear and the front whilst up ahead is another accident damaged loco this time possibly a 4-4-0, whether there was any relevance to the two engines being on the same line is not confirmed, perhaps more likely both have been stored here pending assessment / repair. Engines of this class were spread throughout the system. By 1921 No 2559 was at Didcot but in 1934 had moved north-west to Brecon. For a short time it was then at Oswestry before returning again to Brecon in December 1940 and from where it was withdrawn from GWR service in October 1940 and sold to the Government for ROD service as No 198. For war service the existing ATC equipment was removed and a Westinghouse pump fitted for working air-braked trains. In addition pannier tanks were added together with condensing equipment making its revised appearance as per No 179 below. Livery was also now plain black with the WD arrow symbol. Some of the 100 members of the class that were requisitioned had these and similar modifications carried out at Eastleigh although No 198 was attended to at Swindon. It is not certain what role No 198 played in the post 1940 period as it had not been shipped to France before the German invasion. Post war it was one of several of the class that found its way to China, after which its fate is unknown. *HL04B 088 (85812)*

Opposite bottom: WD No. 179 [ex-GWR No. 2466] was one of ten Dean Goods fitted with pannier tanks and condensing equipment in 1940. Intended for service in France, with the Nazi invasion it remained in the UK. The purpose of the tanks etc was to increase operational range and reduce steam emission which might otherwise betray the locomotive's position to hostile forces, although it is unlikely that this equipment was ever used. This locomotive remained in the UK and from 1940 onward was located in Kent where initially, it probably attended a rail-mounted naval gun used to bombard France. Later known allocations were: East Kent Railway 1943, Southampton 1944, Longmoor Military Railway 1944. The tanks were probably removed before the end of the war; this engine was shipped to China in November 1947 under the United Nations Relief & Rehabilitation Administration's programme and its later fate is unknown. *HL04B 085 (49703)*

Above: ROD 2-8-0 locomotives in the workshops of Armstrong-Whitworth, circa 1927/8. The Robinson 2-8-0 design had been adopted as the standard freight locomotive for the Railway Operating Department in WW1. A total of 521 were ordered by the ROD between 1917 and 1919, the latter order to maintain manufacturing following the European conflict. A total of 311 worked in France after which they returned to the UK where many found employment with a number of railway companies covering for a backlog in overhauls and repairs. Between May and July 1919 the GWR bought 20 in what were basically brand new condition and numbered them 3000 to 3019. At the same time they hired a further 80 although these varied in condition, the majority having previously seen service in France. They were numbered 3020-3099 and 6000-6003. Of these 80, five came direct from the Lancashire & Yorkshire Railway and two from the London & South Western Railway, in both cases being engines previously on loan. Although allocated GWR numbers as stated, it is possible some never actually had plates fitted whilst others may have simultaneously carried ROD and GWR numbers. The 80 engines referred to had a brief spell in traffic before all were withdrawn for assessment. Thirty were considered worthy of an extended life and overhauled at Swindon which included fitting copper fireboxes, they also appeared in GWR green livery. These 'better' engines now took the numbers 3020 to 3049 and returned to traffic in 1926/7. The remaining 50 were given notional repairs only and painted black. These took the numbers 3050-99, the whole episode seeing some considerable renumbering. They too returned to traffic but were withdrawn and scrapped when in need of major repairs. First to be withdrawn was No 3067 in October 1927 and the last No 3093 in December 1931. Of the 'good' engines that had been retained a number lasted in traffic well into the BR era with the last going in 1958. Meanwhile in 1927 the LMS purchased 75 engines it seems primarily for their tenders. A total of thirty were then sold by the LMS to Armstrong-Whitworth who prepared them for export to China as Chinese Government Railway class KD4. *HL 08A (59287)*

The unseen Great Western

Above: 4-4-0 No 4120 *Atbara*, doyen of the class of the same name and built at Swindon in April 1900. This class of 40 engines followed on from the 20 members of the Badminton class built at Swindon in two batches between December 1897 and January 1899. The Badminton class name was a sop to the Duke of Beaufort; and through whose estate at Badminton the GWR were building part of their new line to South Wales. The Atbara class were a development of the Badminton type and built with straight frames. They took their names from contemporary military engagements or senior army commanders. Later engines were named after cities of the British Empire. There were 40 engines in the class. For completeness we should mention the final 20 engines of similar type, the Flower class. These had deeper outside frames than the Atbara type and were also fitted with a new design of bogie based on that carried by the French Atlantics. As built the engine *Atbara* carried the number 3373 but after just six months in traffic the name *Maine* was substituted, this in connection with a special working of 29 November 1900. In February 1901 the name was again changed to that of *Royal Sovereign* in connection with royal workings. In 1912 in connection with a general renumbering of the three 4-4-0 types the Atbara class were renumbered from 4120 onwards, one exception being the former No 3382 which had been withdrawn following accident damage and deemed beyond economical repair. Now as No 4120, it continued to see service on passenger trains although larger engines and longer / heavier trains meant a cascade effect from what had once been front line duties. Here No 4120 is seen on Lapworth troughs, the single headlamp on top of the smokebox indicating a stopping service. A snapshot for 1921 shows the engine allocated to Tyseley. No 4120 was withdrawn from Leamington in September 1929. *HL09D 013 (68732)*

Opposite: Cobham class 2-2-0 No 157 photographed on the troughs at Goring on 25 November 1895. The type were also known as the '157' class, and sometimes referred to as 'Sharpies', the latter as they were replacements for the Sharp 2-2-2's of 1862. In many respects they were similar to the 2-2-2 'Queen' class engines excepting having sandwich outside frames and a brass rim adorning the driving wheel splashers. Ten engines were built with the writers of the RCTS of the opinion that when running in their original condition with domeless boilers and open splashes, they ranked amongst the best looking of all GWR designs - subjective of course. (Closed driving wheel splashers started to appear in the 1880's.) Just three members of the class were named, No 162 appropriately *Cobham*, and Nos 158 *Worcester* and No 163 believed to have carried the name *Beaufort*, both the last two were removed by December 1895. It is remarkable to think that at one time they were rostered to main line express duties, an example being when No 162 took a train of 160 tons from Paddington to Birmingham via Oxford in 2 hrs 12 minutes including two slacks. (The GWR were later to promote their 2hr Paddington - Birmingham expresses on the new and shorter line through Princes Risborough.) The class appear to have spent most of their time divided between Westbourne Park and Wolverhampton (the London depot at Old Oak Common was not opened until 1906). Withdrawals commenced with No 157 in June 1903 and all but one had gone by December 1906. Surprisingly No 165 survived until December 1914 having served out its last days at Oxford. It had also been fitted with ATC in 1906. Mileages for the class were also creditable and varied from 750,000 for No 162 to a staggering 1,228,000 for No 160, this figure achieved in just under 26 years of service and equates to over 47,000 miles per year. The troughs at Goring were brought into use on 1 October 1895 and it will be noted there is still an amount of builder's debris visible. *HL06 001 (92082)*

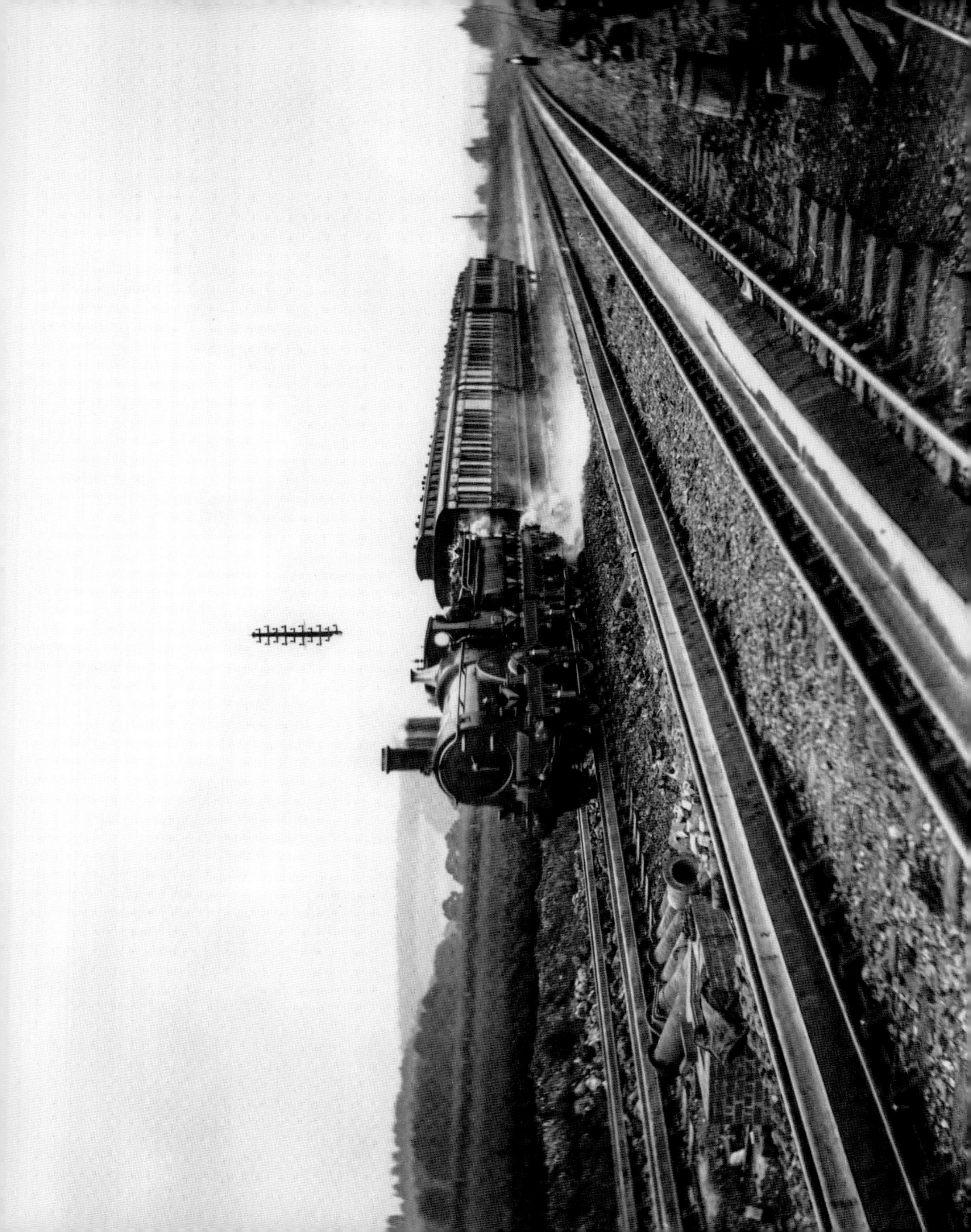

The unseen Great Western

The unseen Great Western

Opposite top: GWR No 1 was an experimental design whose origins go back to 1880. This engine started life as a standard gauge 4-4-0T in 1880 having outside frames and a leading bogie of unusual design so far as the springing is concerned. (The reader is referred to the RCTS Book 6 page F41 for more detail.) In this form it ran until rebuilt as a 2-4-0T in May 1882. At the same time further surgery took place with the side tanks shortened and a conventional copper cap added to the chimney. In its revised form it operated successfully from Bristol on services to Taunton and Salisbury before being moved to Plymouth and then north to Chester. In total it achieved some 530,000 miles during a 44 year career until withdrawn in July 1924, although only 2,870 of this was in its original 4-4-0T form. This was William Dean's first experimental engine, the fact it was a success may have encouraged Mr Dean to try a somewhat radical 4-2-4T which did not work, perhaps not surprisingly being prone to derailment. In the view reproduced No 1 is seen at an unreported location and with a proud crew alongside. Note the smokebox wing plates and warning bell on the cabside; an early form of emergency communication. The dome, safety valve and front axlebox covers positively sparkle although the copper cap to the chimney is somewhat tarnished. Refilling the sandboxes would have involved a hike to the top of the tank with a heavy bucket of sand. *HL038 097 (56934)*

Opposite bottom; One of the ubiquitous 'Metropolitan' tanks (the name often shortened to that of 'Metro' tanks), this one No 1497. In total 140 engines were built, all at Swindon and in the period 1869 through to 1899. In reality they were of three distinct variations, 'small, medium, and large, the differences being in the coupled wheelbase and size of the side tanks. No 1497 seen here dated from August 1892 and is seen with a B4 type boiler and top feed. For many years the class were synonymous with suburban workings out of Paddington with 50 engines fitted with condensing apparatus for working the Metropolitan lines. Other members of the class could be seen at centres such as Birmingham, Bristol and Gloucester again working suburban type services. No 1497 has a proud crew posed on the footplate; note the driver's snap tin on the framing and what is also clearly a red distant signal on the nearby bracket. This engine remained in traffic until November 1944. *HL038 122 (85479)*

Above: French Atlantic 4-4-2 No 102 *La France*, delivered to the GWR from the builder Société Alsacienne de Constructions Mécaniques in October 1903. At Swindon Mr Churchward was considering developing a larger locomotive design and had viewed development in both America and France. On continental Europe some of the best performances were being achieved by the Nord Atlantics, hence the purchase of No 102 and then two further slightly larger engines, Numbered 103 and 104 in 1905. All were four cylinder compounds with two high pressure outside cylinders driving the outside cylinders and front axle and two larger low pressure cylinders driving the second coupled axle. The driving axles were coupled. Whilst all three performed well and were the equal of current Churchward designs there was no significant improvement, although they gave a smoother ride and a reduction in load stress on rods and axleboxes; this due to the divided drive. Whilst being excellent engines for the period, so too were the equivalent designs of Mr Churchward and which therefore sealed the fate of compounding so far as the GWR was concerned. (In 1926 Messrs Stanier and Hawksworth took the idea of a compound 'Castle' to the then Swindon CME Mr Collett but the idea was rejected. Interestingly Mr Stanier did not pursue the idea a few years later when he was in charge on the LMS.) What the French engines did show was an improvement in bogie design and which was subsequently adopted by the GWR, LMS and BR. All three spent their later years based at Oxford. No 102 remained in service until October 1926. Whilst compounding may not have been pursued on the GWR, on the LMS there were plans in 1941 for two compound 4-6-2 locomotives. Apparently Mr Stanier visited Derby on the 5 November 1941 and following discussion specified that the compound 4-6-2 was to have 15inch HP cylinders, 23 inch LP cylinders both with 26 inch stroke. Boiler pressure 300 psi. Tractive effort calculated as 39,400 lb. Subsequent consideration came up with a revised design with 17.35 in x 26 in HP cylinders and 25.5 in x 26 in LP cyls again 300 psi BP. On this the calculated tractive effort was now 48,750 lbs which would have made it a very slippery beast, accordingly a proposal was made for limiting the tractive effort by starting the engine as a 2 cylinder controlled by a special diverting valve. Sadly wartime exigencies precluded further development and the idea was not continued under Stanier's successor. (With thanks to Bob Meanley.) *HL13F 048 (65566)*

53

The unseen Great Western

Above: The new oil gas works at Newport Ebbw Junction on 23 September 1927. This plant produced gas for carriage lighting and cooking in restaurant cars. Prior to 1923 the GWR had but one such plant located at Cardiff whilst at the same time the Taff Vale Railway also had such a plant, again at Cardiff. Before 1927 gas requirements at Newport were dealt with by the use of travelling tanks. By 1928 the former Taff Vale plant was in need of renewal and the GWR decided to built a new oil gas plant at the depot at Ebbw Junction. It was described in some detail in the GWR Magazine for December 1928 with the justification for its construction being that there remained, and indeed would remain for some time, a number of gas-lit vehicles, notwithstanding the spread of electricity for carriage lighting. At the same time gas remained the means of cooking in the company restaurant cars and would continue well into BR days. From this statement we might reasonably conclude that numerous of the local trains in the area remained gas lit. (On 13 October 1928 just before the GWR Magazine article appeared, an accident occurred on the LMS railway at Charfield when a passenger train which had over-run signals collided with a freight train. The gas cylinders of the first four vehicles of the passenger train were punctured in the collision resulting in a fire in which fourteen passengers were killed. The following month in Parliament Colonel England [MP for Heywood and Radcliffe] asked the Minister of Transport whether his attention had been called to the existing and continuing use of gas lighting in railway trains; whether he has made any representations to the railway companies that an electric-lighting system should be substituted; and, if so, with what result? Colonel Ashley (New Forest and Christchurch) responded, 'I am aware that a considerable proportion of railway carriages is still lit by gas. The Board of Trade and the Ministry of Transport have taken the view that the use of electricity for this purpose is desirable in the case of all new and reconstructed passenger stock, and the companies, as a rule, have in recent years ceased to fit ordinary passenger carriages with gas lighting equipment, with the result that a gradual substitution of electricity for gas is taking place. The attention of the companies has also from time to time been drawn to the desirability of using electricity instead of gas, more particularly on main line and express trains, but the risks arising from the use of gas have not been held to warrant its absolute prohibition as an illuminant in railway carriages.') Gas lighting was last used on passenger services on the Hemyock branch from Tiverton Junction. This was due to the short distance involved and the commensurate slow speeds which were insufficient for the dynamo to charge batteries. Newport Ebbw Junction shed was brought into use on 17 July 1915 and was of a similar pattern to Old Oak excepting that just two turntable roundhouses were provided. (Available land at the rear allowed for the provision of an additional two but this was never carried out.) As with Old Oak and indeed some of the larger depots elsewhere a separate covered repair shop was provided, this had 12 roads some directly accessible via sidings and others via an external traverser. This repair shed was not fully functional until after WW1 as initially part had been given over to the manufacture of shells. Much of the work undertaken from the depot was in the form of freight and of course there were numerous pannier tanks allocated throughout its life - see right of image. Possibly the need for two gas plants in close proximity was later found to be unnecessary - more carriage conversions to electric lighting? Indeed the late E Lyons has described the building (1947) as now being for 'boiler washing'. Worth mentioning is that Harry Holcroft who had designed the Stafford Road gas works was later transferred to Swindon where his first task was to design a new gas works there as well. *HE2 020 (11077)*

The unseen Great Western

A rear view of Swindon Station West Signal Box. This box had opened on 27 October 1912 and replaced a timber box of the same name which closed on the same date. As can be seen it was a large structure, 74' 2" long 12' 2" wide and 13' high to the operating floor. To a standard design (although sizes varied according to the location), it was built around a timber frame but had a brick skirt added in WW2 to afford some protection against blast damage. Inside was originally a 163 lever frame with the levers set at 4" centre and with horizontal tappet locking. This frame lasted until 13 July 1932 when it was extended to 174 levers and the locking modified to the 5-bar vertical tappet type. The position of the clock on the exterior was unusual and may simply have been to allow drivers to see the time. A similar timber structure (Swindon Station East Signal Box) existed at the east (London) end of the station and with a smaller frame of 80 levers. East box had been brought into use seven months before west, it too replacing a smaller structure. (Interestingly at the eastern end of Swindon station a replacement mechanical box was erected in the early 1960s with 107 levers, but it was never commissioned. Instead the structure and frame were removed and used at Radyr Junction - relocked as necessary to suit the layout.) Swindon West worked to three signal boxes, the already mentioned Swindon East, west to Rodbourne Lane or as far as Wootton Bassett East if all the other intermediate boxes were switched out of circuit. On the Gloucester line Purton had to be open when trains were running due to the level crossing whilst Loco Yard, despite being shown as fitted with a block switch similarly remained open. Although undated the view is from later BR days as witness the profusion of tubular post signals - note too the rear of the triple disc at the base of the box. With two chimneys from the roof, clearly within the actual box were two stoves, signal boxes not always the cosiest of places due to cold draughts that would come up between the levers. Staffing levels are not reported although it is likely that on early and late turns there would have been two men per shift plus a booking boy. This may well have been reduced to single manning at night. The small lean-to wooden shelter appears to be the signalman's cycle store, possibly with the signal box coal bunker nearby. Trackside a double compound will be noted along with further ground signals. Note too the (frosted) board crossing and the timber walkways over the various signal wires, the latter otherwise a distinct trip hazard. In winter and at locations where traffic was limited at night, the early turn man may well deliberately tread on the wires when he arrived but before entering the box to start duty. This was to try and effect some movement in wires and pulleys that might otherwise have frosted during the night. In the background is of course the locomotive works. Mechanical signalling survived at Swindon until 1968, when on 3 March all the boxes in the area were closed with control given over to an 'NX' ('entry-exit') panel and MAS (multiple aspect signalling). At the same time the track layout was rationalised to suit the then level of traffic and all points motor driven with control exercised from a new building housing the Swindon panel (signal box). As with at mechanical boxes, the panel was modified several times over the course of its life to suit layout changes. It was finally superseded and closed from 20 December 2016 with remote operation now given over to the Thames Valley Signalling Centre at Didcot. Whilst only odd items directly related to the mechanical signalling era at Swindon survive, the 1968-2016 panel was rescued and installed in a purpose built building at the Didcot Railway Centre. *HS3 005 (62148)*

The unseen Great Western

Above: the interior of Ealing (Broadway) signal box. This was a 39 lever frame set at the older 5¼" distance in consequence of the double-twist locking. The interior equipment dates from prior to WW2 and on the block shelf consists four pairs of single deck Spagnoletti instruments, plus nearest the camera two permissive block instruments, these are for the goods line to Acton West. A large number of signals are power worked hence the short levers in the frame. For ease of recognition to the signalman, signal levers applicable to relief line signals are painted with a black lower half, the top being the conventional red or yellow. A replacement 65 lever frame at 4" centres was installed and operational from 7 October 1945. The box closed from 20 March 1955. *HS3 008 (61251)*

The unseen Great Western

Opposite bottom: Interior of Old Oak Common East, another timber box of similar external style to Swindon and also one where brickwork was added to the lower half in WW2. Opened on an unreported date in 1906 it replaced an earlier box at West London Junction East. (The name change related to the opening of the new locomotive depot.) As at Swindon it too was expanded, from the original 131 levers at 5¼" distance to 160 levers at 4" centres from 26 April 1927. It closed from 8 October 1962. The view is likely post 1927 as the light coloured levers reversed in the frame are distant signals with the levers painted yellow. Before this when red distant signals were common the corresponding lever was painted green. (The colouring harked back to very early days with the rhyme, 'White is right, red is wrong, green means gently go along'.) Two men and a boy are employed, the nearest man operating one of the Spagnoletti block instruments (double-deck this time) and at the same time tapping out a bell code. The booking boy would quickly have to learn the tone for each bell and recognise the code received; often sent in an almost continuous sound rather than with pauses in-between. The striped levers are for placing detonators and have ovoid lever lead plates. Note the highly polished lever tops and that the levers all have brass lead plates. A hand-lamp stands ready on the block shelf; if needed quickly this would be to give a red 'stop/danger' warning so the glass would always be turned to red, 'just in case'. Other instruments are located on the window sill for the simple reason there is no room on the block shelf; that nearest the camera probably a lamp repeater of some sort - which would ring if the paraffin lamp on one of a series of signals went out. Illuminating the frame are various electric lamps with the shades directed towards the track diagrams, meanwhile older oil lamps are suspended from the ceiling in case of a power cut. *HS3 009 (61252)*

Above: A much smaller signal box, this one at Drayton Green on the West Ealing loop. To some signalling is almost regarded as a 'black art', yet the principles are extremely simple with just one train permitted on a single section of line from 'A to B' at any time. Add a few points and the necessary levers and locks and then a few signals, and repeat for the opposite direction and it can be seen how it might then appear to be confusing. Here the basics are apparent. The levers are mostly laid out to correspond with the sighting of the actual equipment on the ground. Hence at each end of the frame is a lever for a distant signal then the stop signals and the various points and controlling discs in the middle. This was a frame of 23 levers, the block shelf containing the bells and instruments by which the signalmen communicated with each other. Again subsidiary items are placed on the window sill. The two round handles are for slackening or tightening the wires to actual signals to accommodate expansion and contraction. It is amazing how quickly a signal wire can go slack in direct sun and how quickly it becomes taught in the cold. Black and white striped levers are again detonators. (Later these would be identified further with black and white chevrons; pointing up for the up line, and down for the down line.) Short levers 15 and 16 are switch levers painted red with white bands; the box could switch to either West Ealing or Hanwell. Brass plates below the instrument or bell describe the purpose whilst the circular item above 15/16 is a lever collar, to be placed on a lever as a reminder it was not to be pulled; they also made for excellent egg cups! *HS3 053 (69297)*

The unseen Great Western

Above: Not a signal box, although the building design was certainly of similar style as was used for numerous signal boxes, it is instead the 'Hump Ground Frame' at Banbury hump yard. For many years Banbury was a vital exchange point for goods traffic from the GWR and on and off the LNER via the connection to Woodford Halse. Accordingly a hump yard was provided east of the four running lines with reception sidings to the north and sorting sidings to the south. The facilities came into use in 1931 and were essential during WW2. A number of running signals were controlled from the box and for this purpose a conventional forward facing 9 lever frame was provided. Originally this was a McKenzie and Holland type but a replacement 5-bar locking was installed in 1942. As the name Ground Frame implies, this was not a block post although the name Ground Frame should not be taken literally as the frame was elevated to the usual level operating floor. In addition to the lever frame there was a describer panel of 21 buttons operating the points into the sorting sidings on the south side - to the right of the image. The photograph shows a single wagon 'cut' from the main rake being pushed towards the hump and will now run into its correct siding. Speed was controlled by the shunter running alongside and applying the wagon brakes, possibly with the aid of his brake stick, to impede progress. The intention being for the wagon to just kiss the buffers of other vehicles waiting in the same siding. Too much speed and a veritable crash would occur, shocking the contents of both wagons; truly excess speed would result in a derailment. Too little speed and the wagon would stop short meaning the shunting engine would be required to push all the wagons together. For the men on the ground shunting was a hazardous operation especially at night or in poor visibility. Total wagon capacity in the 20 hump sidings was 1,477 wagons with an additional 389 in the six reception sidings. The GWR/ BR(W) had only a few hump yards, others being at Severn Tunnel Junction, Bristol East Depot, Rogerstone and in BR days Margam. At the last named technology in the form of retarders was installed meaning the progress of a wagon could be mechanically retarded as it passed over the hump. Operating a hump yard was a skilled operation. The hump at Banbury came into use in July 1931 with this view dating from October 1937. Notice the non standard windows at the south end and also the lighter coloured brickwork underneath the windows, almost implying a longer and larger nameplate had once been provided. The single line in the foreground was bi-directional with access to both the Up and Down relief lines. The train being pushed towards the hump consists of a 57xx pannier tank with shunters truck immediately behind. Ahead is a Sully & Co private owner wagons, another PO wagon from the Stroud Gas Light & Coke Co., a common user North Eastern (LNER) open and a further Stroud Gas Light & Coke Co. vehicle. At least some of these would appear to be loaded with coal. The vehicle on the way down is owned by E Foster & Co. of London. The hump box closed in August 1971. *HS4 005 (64533)*

Opposite: When coal was King; the coal marshalling sidings adjoining Roath Dock sidings at Cardiff. This was a group of over 20 sidings where coal for export might be stored. Unusual for what was probably an official view, the image almost appears to have been taken with a telephoto lens. Two engines are visible, both probably '850' saddle tanks. The area here was also known as 'The Prairie' and was surrounded on two sides by water. The two signal boxes are (nearest) 'Dock Storage South', and 'Roath Branch South' in the background. *HD1 082 (50198)*

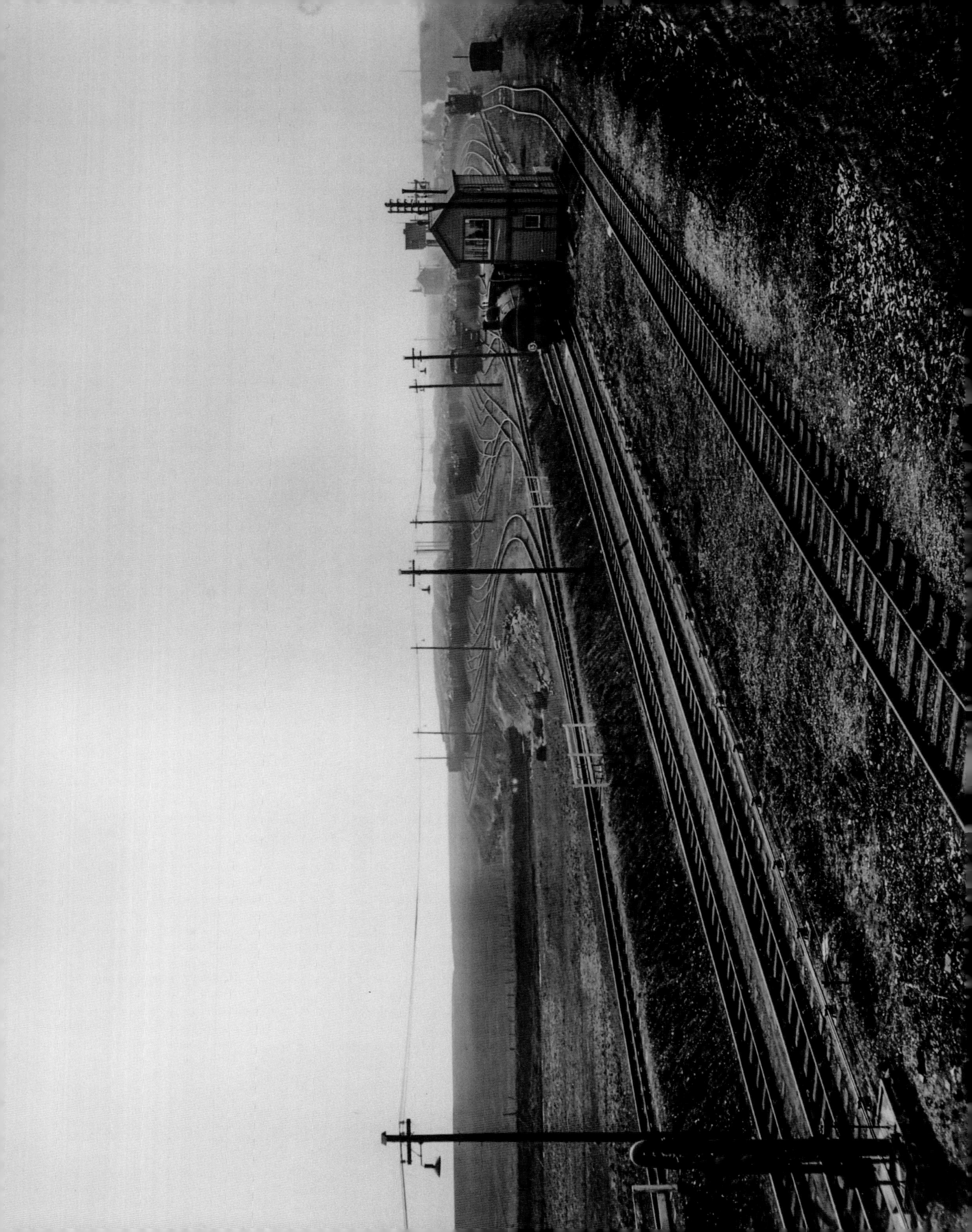

The unseen Great Western

The unseen Great Western

Opposite top: Old coach, new use. This a former 4-compartment 1st/ 2nd Composite similar to but not completely identical to Diagram U4*. The difference being a slight variation in the length; a U4 being just over 28' in length and the vehicle seen marked on the solebar 29'11". Vehicles of this general type were built for branch and secondary service in batches from the 1880s through to the early years of the 20th century. Note it was originally a composite first and second class coach, the first class compartments being the two in the centre. Four (and six) wheel coaches slowly gave way to cascaded bogie stock in the 1920s and beyond, some survivors turned over to departmental use and a few converted into Camping Coaches. We see here No W32 freshly turned out in BR days for departmental use as a sleeping and mess van for the electrical section (New Works). In connection with this conversion two compartment doors have been sealed meaning the interior will now be 'open'. The doors that remain have had extra footsteps added to allow access from ground level - a bit pointless at the end nearest the camera where the door is permanently closed. *(A search has failed to identify an alternative Diagram Number for this coach. The length on the solebar might even refer to that over buffers and the conclusion is it is a U4. Note that several of this type were converted to Mess/ Accommodation duties in the 20s and 30s number series. *HC1 018 (48951)*

Opposite bottom: This vehicle is a Diagram F1 tri-composite double slip coach with compartment configuration:- guard/ 3rd Class/ 2nd Class/ 1st Class/ 1st Class/ 2nd Class/ 3rd Class/ Guard. Six vehicles of this type were built, two in 1880 and a further four in 1882 as 'convertibles' (a narrow gauge body on a broad gauge chassis). The vehicle depicted is possibly No 1019 of the second batch. As built these vehicles were rigid 8-wheelers with the wheelsets at each end 7' apart. They were later converted to conventional 8-wheel vehicles having a Dean design 6' 4" swivelling bogie at each end. Slip coach working had started in February 1858 with the London, Brighton & South Coast Railway and the GWR was not far behind in November of the same year. Slipping on what was now the Western Region continued until 1960 which was also the end of slip coach working generally. The alternative is a Diagram F10 tri-composite double slip of 1897 of which four were built. The open door trackside suggests the view was taken in the centre road of a station but without further clues as to the location. Note the destination board 'To South Eastern & Chatham Railway', which suggests a through working and from the company name prior to 1923. There were some regular through workings between the SECR at Folkestone and elsewhere including as far afield as Liverpool. It appears alternate company stock was used on these trains with engine changeovers at suitable locations; so far as the GWR was concerned this included Reading. There was also a pre-WW1 through service from the GWR into London Victoria. *CN4 053 (59082)*

Above: An unusual vehicle, one of seven (six brake thirds, and one all third) built by the GWR in March 1939 especially to comply to the restricted loading gauge of the Burry Port and Gwendraeth Valley Railway. Up to that time 4-wheel coaches had been used and this was the first time bogie vehicles had been used. (The brake thirds were 57' in length and the all third slightly shorter at just over 55'.) They were all 18" lower than a normal coach and 3" narrower. Their life on the BPGVR was limited as the final sections of this line closed to passengers in 1953. The carriages however found new uses including services to Tredegar whilst three were hired to the Royal Navy for working to Trecwn. One of the RN trio was overhauled at Swindon as late as 1963. The same RN vehicles found their way to be filmed at Bath Green Park in 1965 in company with an LMS 3F 0-6-0T. In the view above, No 1323, the lowest number of the six brake third vehicles, is seen attached to a 'Cordon' (gas tank vehicle). The designation applicable to their working is painted on the coach end. Despite the presence of end windows the vehicles were not auto fitted. *CN4 049 (59078)*

ABERYSTWYTH AND MANCHESTER (LONDON ROAD)
FIRST
306
FIRST
THIRD
THIRD
THIRD
THIRD
306
THIRD

The unseen Great Western

Opposite top: Three car GWR diesel railcar, comprising twin-set Nos 35 and 36 and an intermediate 10-compartment third which has been through wired. A through corridor was available between the vehicles. The GWR built two pairs of twin-set railcars Nos 35/36 and 37/ 38 which entered service in 1941. Seating in the two cars amounted to 104 seats which included a small buffet able to provide hot drinks and light snacks to travellers. Whilst the engines were diesel powered heating came from a steam boiler. As with the earlier single car railcars they were an immediate success and saw service on cross country routes including Cardiff - Birmingham. The popularity of the single diesel cars on this route pre-war had seen steam hauled trains often having to replace the diesel and in light of the slower schedules of wartime yet with continuing high passenger numbers, in 1941 a trial was made with the addition of an intermediate coach so increasing the seating capacity to 184. This was a success and at least one of the pairs if not both regularly had a 70' 10-compartment corridor third assigned (Lot 1337, all third to Diagram C46 of 1925). Through control wiring between the two end vehicles was achieved with cabling under the step boards of the intermediate vehicle. Both sets operated successfully during WW2 but car No 37 was destroyed by fire in 1947. It was replaced by No 33 rebuilt as a single ended vehicle with a corridor at the opposite end. The GWR railcars showed the way multiple unit trains would develop under BR with both pairs surviving well into BR days although in the final years reverting to simple twin-sets. Further views relative to the GWR diesel railcars appear on pages 69 and 70. *HC2 060 (53143)*

Opposite bottom: Cambrian Railways composite coach No 306. This is one of four similar vehicles owned by the Cambrian Railways, each having two first class and five third class compartments. Lavatories were also provided at each end. Under Cambrian ownership they were numbered 306/7/10/11. All were 54' 6" in length with the first pair built in 1905 and the second pair the following year. On paper all four passed to GWR ownership but in reality there were just three vehicles, Cambrian No 310 having been wrecked in the 1921 Abermule disaster but not officially shown as withdrawn until 31 October 1925. The other three vehicles, Nos 306/7/11 took GWR numbers 6327/8/30. Renumbering started with No 307 on 11 February 1923, followed by No 311 on 26 August 1923 and finally No 306 on 12 January 1924. All three provided over two decades service to the GWR Nos 307 and 311 withdrawn in 1946 and 1947 respectively, whilst No 306 continued working into BR days lasting until February 1954. Overall they were handsome vehicles originally painted in bronze green later altered to standard GWR colours. *HC3 019 (53905)*

Above: This is a most unusual view showing one of the vehicles from the 1942 'ALIVE' train ('Alive' was the code name for this GWR train) assembled for 'S.H.A.E.F.' (Supreme Headquarters Allied Expeditionary Force) in WW2. The vehicle is (or was) a standard corridor Collett third to Lot 1623, Diagram C77 delivered in March 1940 as one of a batch of 95 identical carriages taking the numbers 501-95. Dimensionally they were 60 11¼' x 8' 11". At some point post 1942 it was given armour plating and bullet proof glass plus window shutters and covers over the roof ventilators. Livery might appear to be all over brown but the respected historian Tim Bryan suggests this could have been khaki and later olive green. The additional protection added to the weight with the numbers on the end showing '38' tons compared to the usual 31 tons. The vehicle formed part of a train for the use of Winston Churchill and Generals Eisenhower and Montgomery plus staff. After D-Day the vehicles were dual fitted with the Westinghouse braking system with the train seeing service in Europe. For information by 1945 the train formation consisted of 'Monster' vans, a generator van, brake thirds, No 574, a restaurant car, conference car, two first class sleepers and a third class sleeper. Some of the vehicles were also modified internally to suit their new purpose. Out of the train, No 574 was the only vehicle to receive armour plating and which from close study of the photograph also appears to include some extra sheeting to the roof. Post war all the vehicles involved were returned to ordinary traffic, No 574 was restored to conventional condition in June 1946. *HC1 093 (53059)*

The unseen Great Western

Above: 'Emergency Train No 1' - but not an Ambulance train; the Red Cross is not depicted. Instead this is a 'mobile control' train, one of two on the GWR with the other companies also preparing similar. Their purpose was to take charge of movements and operations during emergency situations. The train would be taken to a suitable location away from the raid or incident, perhaps a siding in the middle of nowhere, and connections made to the telegraph wires. From here directions would be given such as to prioritise certain types of traffic. Exactly where the two trains were normally stabled, certainly not at the same location, and/ or how often they might have been used is not recorded. As with Ambulance and trains for VIPs, the train was self contained and would include telephones, route maps, etc. Catering was also an important issue and as the last passenger vehicle we have an H38 Collett composite diner, ten of which were built to Lot 1451 in 1930, Nos 9601-10. The identity of this particular vehicle is not known. Livery of the main train would appear to be overall brown or possibly even black. The white painted water tank at the rear, Diagram DD1 and No 30 dates from the period 1886-1900. The more usual purpose of these tanks was being sent to a specific location where drinking water as required rather than being 'mobile'. For example in WW1 the GWR sent a number of water tanks to Worthy Down north of Winchester to provide drinking water until a permanent local supply could be established at what was a new army camp. The GWR chemist at Swindon would also receive regular samples of water from country stations where water was drawn from a well or local river. Before the expansion of the mains water network and in other areas local filtering and purification, pollutants might be present which would then trigger the despatch of a drinking water tank until the issue had been resolved. The vehicles seen were returned to ordinary use after WW2 with the interior of the Restaurant Car modernised by Messrs Hamptons. (BR also had their own civil defence /emergency trains prepared in the 1950s. One remained under cover at Craven Arms for many years whilst others were located at varying locations including Tunbridge Wells West and Doncaster.) Here the location is Swindon West, see also view on page 55. *HC4 029 (59058)*

Opposite top: New use for old. A former 'Bars 1' 57' Toplight Brake tri-composite (later modified to composite) No 7546 of February 1907 and one of 15 vehicles built to diagram E83, Lot 1145. The vehicle was photographed on 31 July 1957 having been converted to a departmental messing and sleeping car No DW150124, for use at No 6 shop Stafford Road Works. Note only two doors remain accessible and the corridor connection has been removed. It is not known how long it remained at that location or indeed in such use for it was condemned in November 1968 and scrapped at Bird's Long Marston in February 1969. A further, unidentified conversion, appears in the background. *HC5 016 (69358)*

Opposite bottom: A second departmental conversion, formerly an all-third 'Bars-1' 57' Toplight of March 1908 number 3643, Lot 1136, Diagram C28 and one of thirty identical vehicles. This vehicle was converted for departmental use in March 1956, in this case a Gas Works staff riding van and renumbered DW150042. It may later have been used as a dormitory coach. Again the corridor connection has been removed and in this case the toplight windows sheeted over. Once again there is limited door access and it may be noted a handbrake has been fitted. Livery of these vehicles was all-over black. It was condemned in May 1969 and broken up at Swindon in September of the same year. *HC5 016 (69360)*

The unseen Great Western

This photo was taken on 14 June 1927, and shows passenger stock from the Taff Vale Railway that had been inherited by the Great Western Railway at the grouping. As is known, the GWR absorbed a number of railway companies as a result of the grouping which included the taking over of their locomotives and rolling stock. Swindon certainly did not have the resources to replace all of this with equipment of their own designs, even if they had wanted to and in consequence it was a gradual process that saw locomotives called to Swindon and either repaired, 'Swindonised' or scrapped. A similar policy was followed with rolling stock such as seen here at Foss Cross on the former Midland & South Western line between Andoversford and Swindon, with coaches awaiting assessment by Swindon GWR. The MSWJR line immediately north of Foss Cross station ran in a cutting through a gravel bed. Widening the cutting provided the MSWJR with ballast for track maintenance. The gravel was not of the best quality, as it soon crumbled to dust, but it had the advantages that it was cheap and readily available. Two long sidings were laid down to the north of the station for loading wagons with gravel quarried from the side of the cutting. After the MSWJR was taken over by the GWR in 1923, the GWR preferred to use better-quality ballast, so obtaining gravel from the cutting sides was discontinued, and the quarry sidings became redundant. The sidings were therefore then available for storing surplus rolling stock. At the grouping, the GWR had inherited a motley collection of passenger stock from various railways in South Wales. Where there was any doubt as to their suitability for future service, they were usually sent to Swindon for evaluation, and a decision as to whether they could be put to further use, sold or scrapped. However the sidings at Swindon soon became full. Stock therefore had to be parked elsewhere, and the long sidings at Foss Cross were one of the sites apparently seen as a suitable alternative. The vehicles shown in this view all came from the Taff Vale Railway. From right to left the first five individual coaches can be identified as GWR Nos. 1439, 1778, 3864, 2533, and almost certainly 1265, ex-TVR Nos. 129, 295, 310, 46M, and 53M. There would appear to be more 'M' series coaches beyond, after which they become too far away to be identified. 'M' series coaches were second-hand, having originally been Metropolitan District and LT&SR joint stock. *HC1 055 (53021)*

Recorded on the same day, this shows a GWR permanent way brake van coupled behind a wide variety of old passenger stock and again at Foss Cross on the former Midland & South Western line between Andoversford and Swindon. The p/way brake nearest the camera is possibly No.60755 (last digit hidden behind framing). They too are awaiting review from Swindon. The p/way brake nearest the camera is possibly No.60755 (last digit hidden behind framing). There could have been two reasons for the presence of this van. Other photos taken at Foss Cross on 27 June 1927 show that the siding contained a wide variety of stock from various South Wales railways. If the carriages included Rhymney Railway stock, they could not have used the vacuum brakes as they were instead Westinghouse fitted, hence the presence of the van for stability when stabled. The second alternative is it may have been considered unwise to rely on the vacuum brakes anyway, on stock that was clearly somewhat decrepit. The carriage next to the permanent way brake was noted as GWR 4-wheel 5-compartment No.685. Unfortunately no details have been found of the other carriages in the train, and the image is not clear enough to read any numbers. *HC1 055 (53021)*

The unseen Great Western

Above: Interior of a 70' steam rail motor. Beyond the bay of seats immediately nearest the camera, the remaining passenger accommodation comprised 'walkover' seats of similar type to those seen in contemporary trams. These seats had backs that could be slanted either way meaning a passenger might face forward or backwards to the direction of travel according to their wishes. They proved to be very popular. Total seating was for 64 persons and with a separate smoking compartment beyond the door at the end. The notice above this door reads 'Passengers are requested not to smoke or spit in the car', almost one of the few occasions in the period when non-smoking accommodation was prioritised over smoking. The reference to spitting was relative to the spread of TB. Notice too the slatted wooden floor. As is known there were also shorter 57' steam rail motors with a total seating capacity of 40 persons and of similar interior design. Considering that these cars would not always have run full it is commendable that the GWR found it profitable to operate such vehicles which required a permanent crew of three plus the usual overheads of train operation. The story of the GWR Steam Rail motors is told in great detail in the book by John Lewis. *HC1 004 (48937)*

Opposite top: An interesting view, at first thought to relate to the diesel rail cars but it is in fact from 1905 and shows two underframes from steam rail motors - not the 'match-board' type. The frame ends will be noted to have become buffer locked which perhaps implies these are 70' vehicles; a further move in either direction will likely result in a derailment. The location is not known but is assumed it is around the Swindon complex notwithstanding the light track work seen in at least one of the sidings. Otherwise clerestory stock, including a Dean 5-compartment brake third No 2240 from 1890, abounds together with a carriage truck. Note too the unusual vertical point lever at 90° to the track together with that staple of the GWR a line of spear fencing. (Assuming the number 2240 has been read correctly, this coach lasted in traffic until 1935. It had been one of 14 identical coaches built in two batches, 10 in 1890 and a further four in 1891.) *HC2 059 (53942)*

Opposite bottom: Craftsmanship at work in the body of one of the 'razor-edge' diesel rail cars; from the batch Nos 19-34 built on a steel underframe. Approval for the construction of these vehicles together with a further four (the twin sets, page 62) was given in 1938. Whilst we were familiar with two vehicles as the twin sets, the controls of these new builds were electro-pneumatic meaning two double ended cars could be coupled together and controlled by a single driver, how often this may have occurred is not reported. Standard draw gear has been fitted with the whole later covered in metal sheeting. *HC9 056 (53139)*

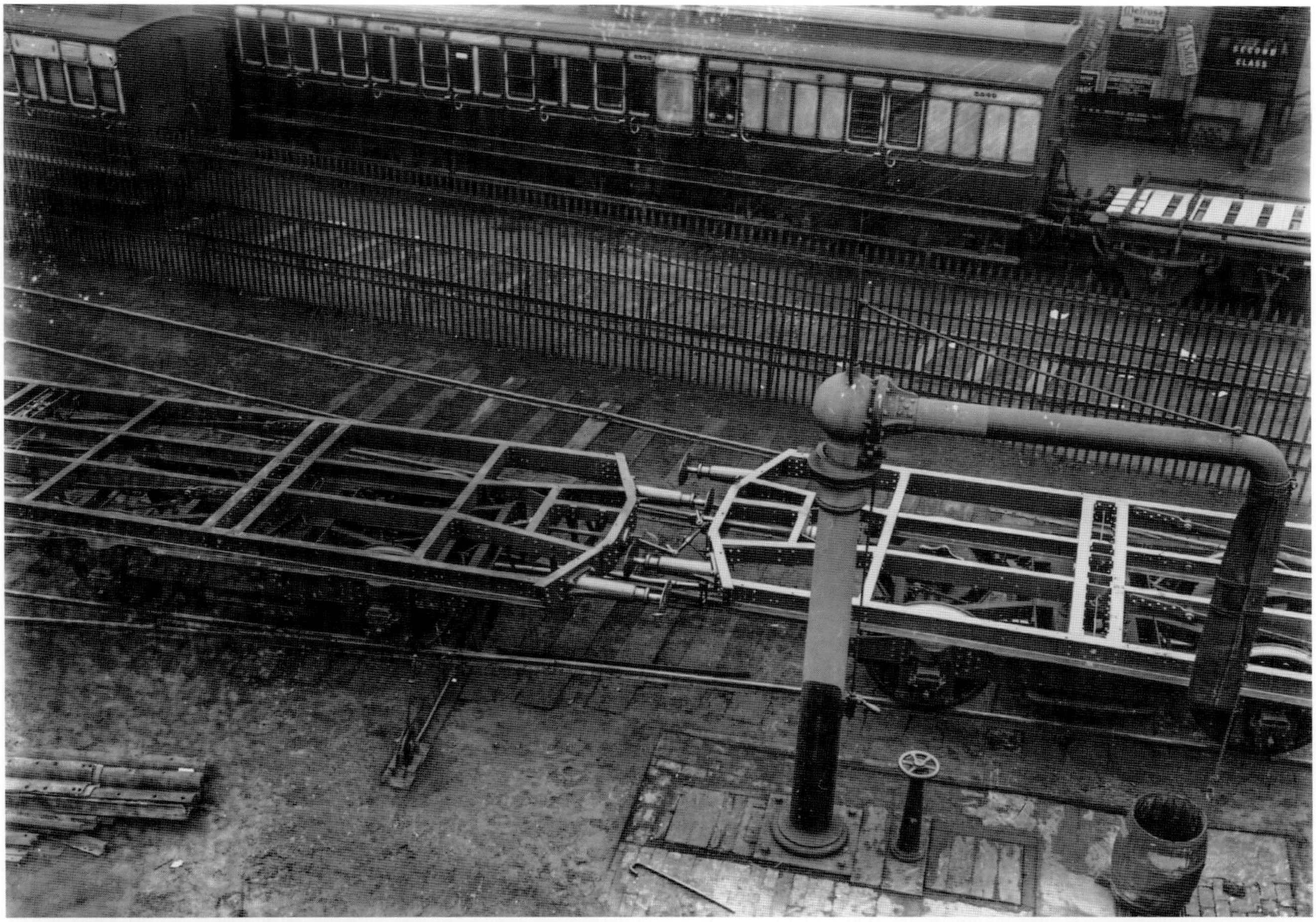
SECOND
CLASS

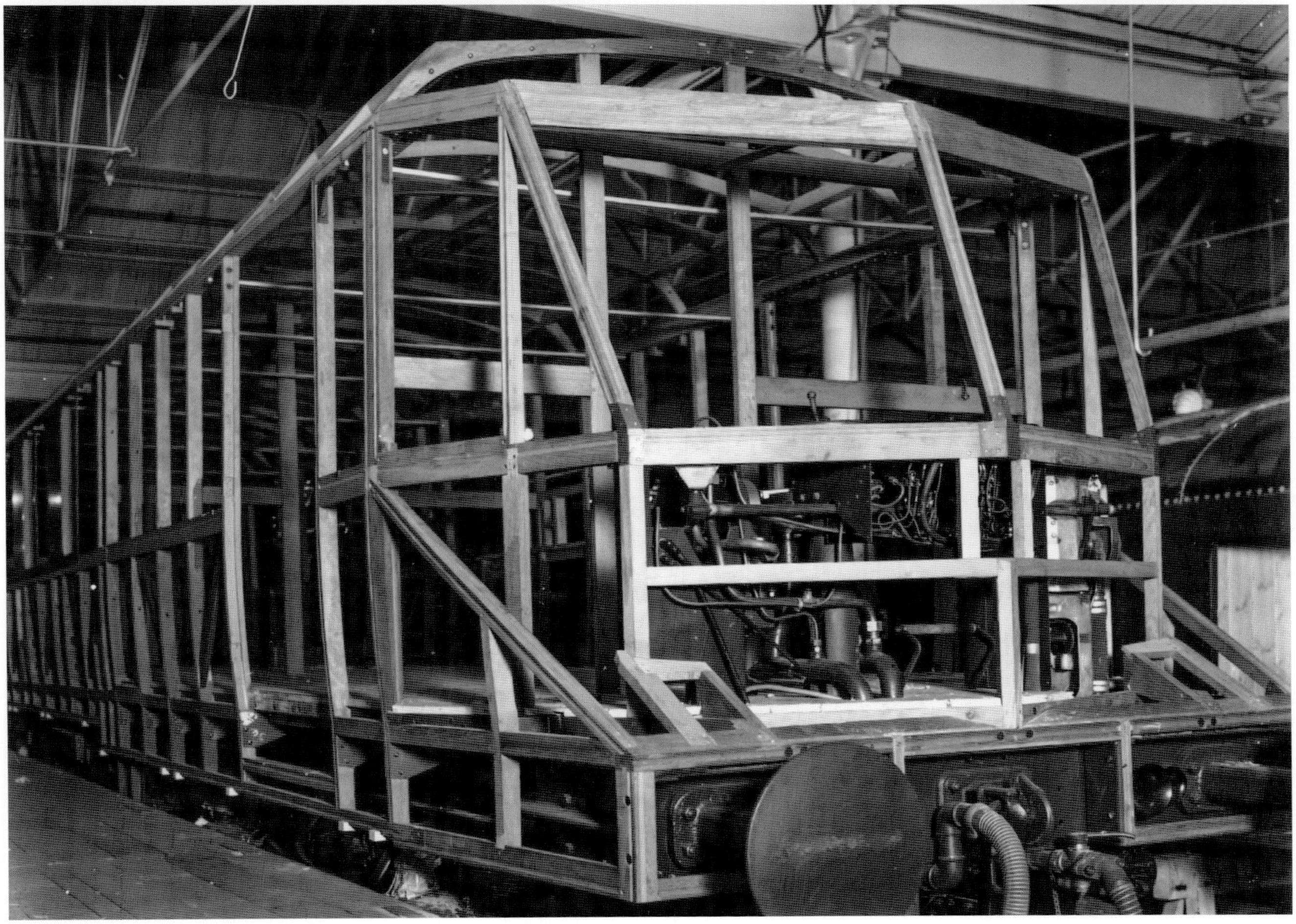

The unseen Great Western

Buffet facilities within a GWR diesel rail car. This style of refreshment area was included in cars Nos 2, 3, and 4 but with the result that the seating capacity was reduced from 69 in car No 1 to just 44 in these vehicles. Buffet facilities were also provided in the twin-set vehicles. *HD3 003 (53069)*

The unseen Great Western

As has been said several times elsewhere, the GWR was very much a self-contained and similarly a self-supporting enterprise. Setting aside day to day traffic for a moment, and instead concentrating on servicing the regular working of the railway, we have to consider wear, tear and repair in everything from locomotives and rolling stock, to infrastructure and operational items. For the first two mentioned, locomotives and rolling stock were either built new at Swindon and previously at Wolverhampton and also the workshops of the absorbed companies, or they were repaired at these locations; repairs other than complete overhauls undertaken at several of the major depots with running repairs a regular feature of depot work. (For the purpose of this discourse we are deliberately excluding the limited number, compared with Swindon, of engines and rolling stock built by outside contractors.) Raw materials, steel, timber, copper, brass and timber were similarly delivered to Swindon to be cut, forged and cast into new or replacement items. We should not forget how the GWR looked after its workforce at Swindon; a mechanics institute and hospital, the GWR staff health scheme the model on which the National Health Service was founded. Is it any wonder the major NHS hospital at Swindon is called the Great Western Hospital and has within its entrance hall a mural and credit to its origins with the GWR. Then there was staff accommodation, the houses in the original 'Swindon village', now rightly listed whilst elsewhere the keen observer will still spot rows of former railway cottages and what were formerly station master's houses up and down the country, sometimes where a railway has ceased to exist for many decades.

Main and country stations whilst perhaps set apart from Swindon by some miles were not isolated so far as being able to obtain supplies of consumables and replacement items, all of which were contained within the massive stores building that once stood proud on the east side of the Gloucester line within the works complex. Whilst Swindon was undoubtedly the largest of the GWR stores warehouses, other major stores locations were at Wolverhampton, Worcester, Reading, Westbourne Park, and Port Talbot. Surprisingly there does not appear to have been a major stores listed for the West Country. By holding quantities of regularly used material it was possible for the GWR to obtain discount on the basis of bulk purchase. (The modern day analogy would be the methods used to purchase goods by the big supermarkets.)

To return to the GWR and the Swindon stores in particular, everyday and consumable items were delivered usually on a weekly basis to major stations which would then be responsible for disseminating these to outstations as required. Some items were swapped on a like-for-like basis meaning for example if a hand lamp were broken, the defective lamp would have to be swapped for a new one; the returned item then hopefully repaired in the tinsmiths shop at Swindon.

In these two views we see Western Region travelling stores van No DW150006, a former Bar 1 Toplight brake, being loaded at Swindon Stores Platform on 24 January 1958 before setting out on what would be one of its regular circuit workings. External livery may have been all-over red. *Top: HC5 034 (69376), bottom: HC5 035 (69377)*

The unseen Great Western

The unseen Great Western

The process of ordering items was via a printed list which would be sent to Swindon and the requested items delivered the following week. Naturally some spares were held locally, one example being say a shunter's pole with smaller stations say having two on hand at all times. Uniforms was another item that might be requested, this time the request going through to the clothing stores. Author Adrian Vaughan recalls how his first uniform issue was a deliberate leg-pull by the staff with items delivered that were totally the wrong size. No doubt he was not alone in experiencing such a induction. Considering how all this was achieved without the aid of the modern computer is nothing short of incredible and yet, similar armies of clerks, filing cabinets and card indexes existed in every business at the time. In the top view opposite, items from the stores are being loaded into the coach for despatch, note too the racking on the inside of the coach against a sealed door. *HC5 036 (69378)* The lower view opposite provides a view inside the stores coach and which inside has been also been modified with heating now from the twin pipes suspended from roof level. Recall this was a guards vehicle and would therefore be occupied during the journey. Among the items carried that may be discerned are of course the box of detergent (the name still familiar today), hand and tail lamps, what could well be a can for carrying sand, together with a drum of some sort plus numerous parcels and packets. *HC5 037 (69379)*

Above: Seen from the opposite end interior the same array of items is seen. The contents of the two large boxes is unknown. Station furniture might also be transported this way in the event of damage requiring an item be returned for the attention of the carpentry shop. Some stores were more urgent than others, The Hotels department had their own requirements, fresh produce being delivered direct to the necessary starting or re-victualling station with other items such as linen dealt with at Swindon. Normal distribution of non-perishable items for the Hotels department was by goods train but urgent supplies were consigned by passenger service. The respected GWR author John Lewis in his article on stores traffic, refers to the Hotels department also maintaining a fleet of two ex passenger brake vans for their own purposes but without explanation as to how or what these might have been used for. With the number of main out-station stores depots to be serviced from Swindon a timetable of deliveries was arranged which necessitated the use of five travelling stores vans. An undated internal GWR publication refers to these five stores vans being in circulation at any one time and covering, South Wales, West of England, Northern District, London & Midland Districts and the Central Wales Division. There is no evidence to suggest any new built vehicles were constructed especially for stores use and instead superseded stock as per the example of the Toplight seen may well have been converted. Certainly we know some former Dean 40' passenger brake vans were used, almost certainly racked out to carry different types of material. One unusual vehicle latterly used for stores traffic was the former royal train brake van No 1069, immediately identified because of its royal clerestory (sloping down at each end). This had been converted for stores use in 1932. We have mentioned the former Toplight Brake No DW150006 previously but worth mentioning is that another Bars 1 Toplight, ex 2375, was converted for stores use in 1955 and numbered DW150005; it was painted all-over red. *HC5 038 (69380)*

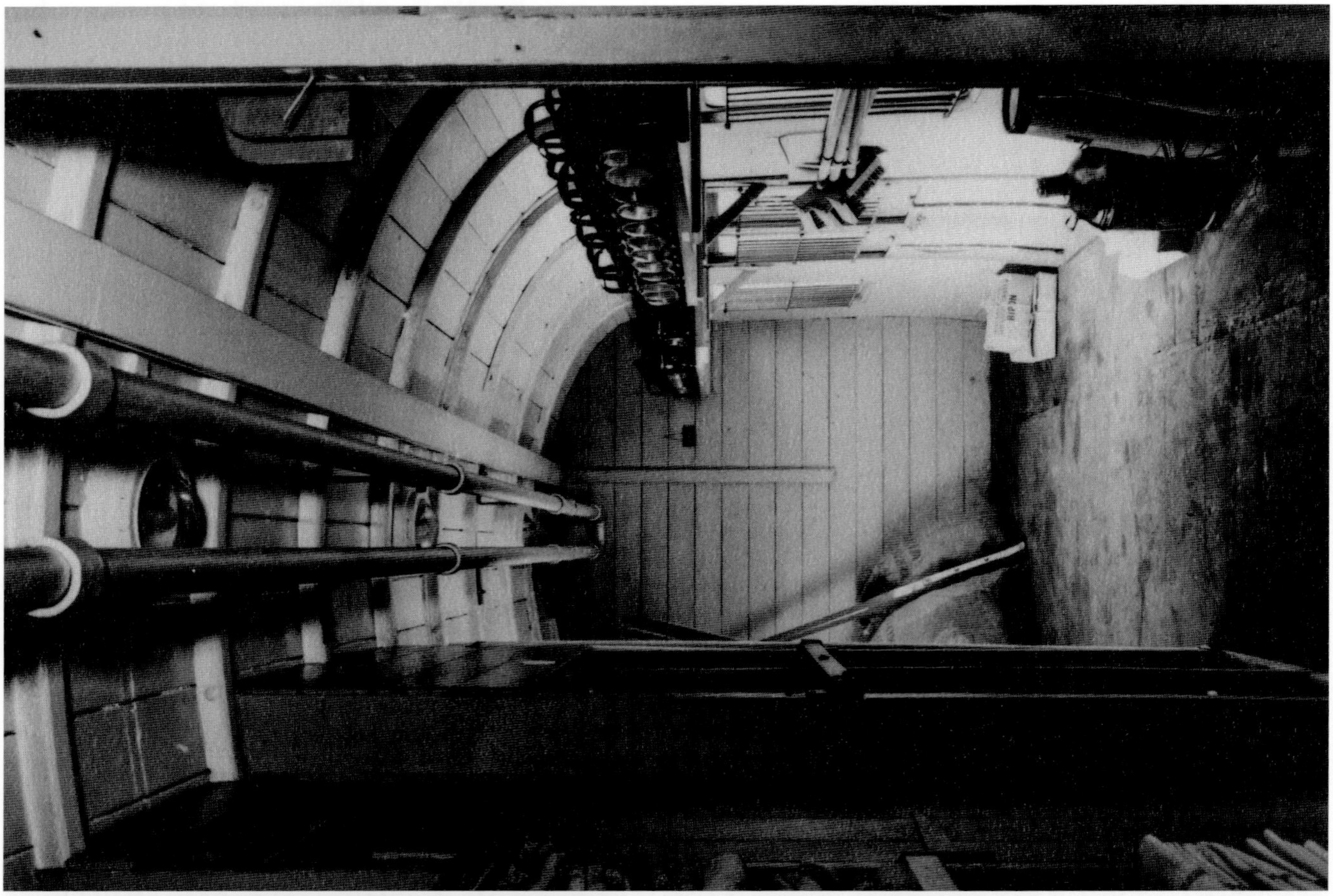

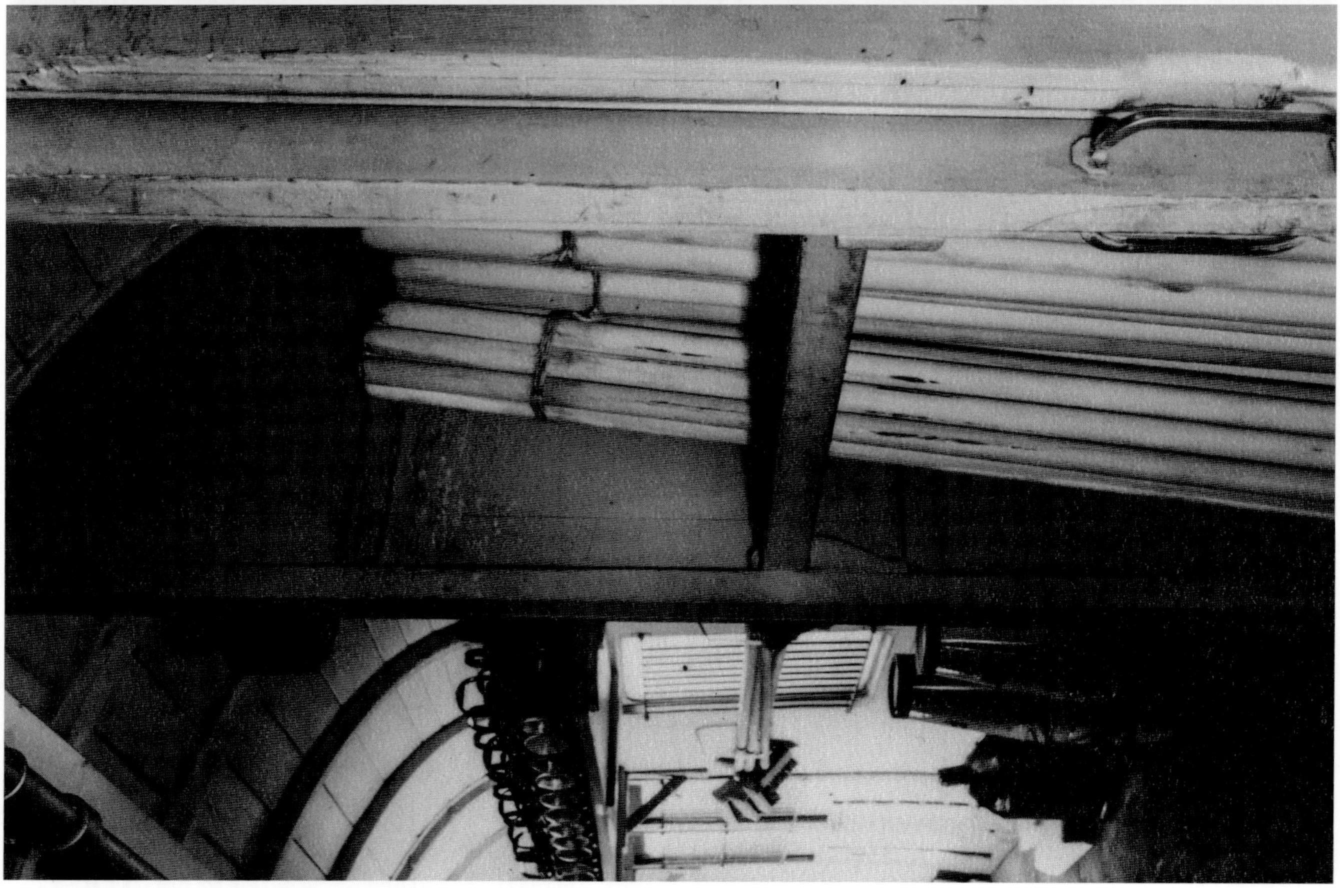

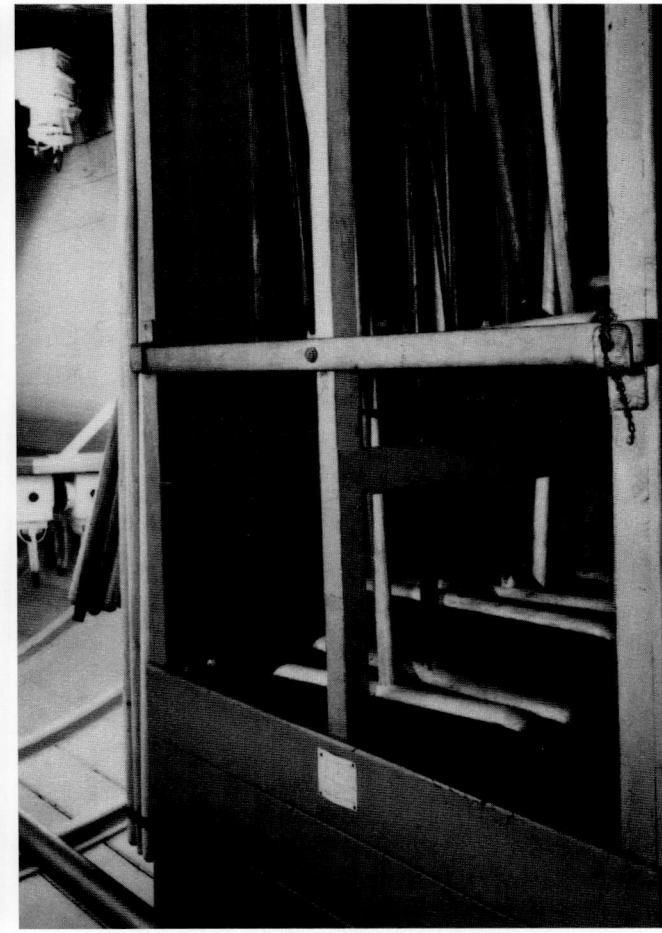

The unseen Great Western

The four images opposite page and above' were also all taken on the same day, 24 January 1958. They show either the interior of the same stores van, DW150006 or another stores vehicle.
Opposite top: Referred to as 'clean towels, dusters, laundry etc'. *HC5 039 (69381)*
Opposite bottom: 'New shunting poles, lamps, firebuckets etc.'. *HC5 040 (69382)*
Above left: 'Clean towels, sponges, dusters'. *HC5 041 (69383)*
Top right: 'New brooms'. *HC5 042 (69384)*

Right: Another interesting view and almost certainly taken to highlight the leaking barrel. A study of the barrels indicates at least two types of oil are contained within the store - the location sadly not given but from the number of barrels clearly this was a major distribution point so we may consider Swindon. The two oil types identified are '*Mo*tor *Lubric*ating Oil' (italics are used for those letters where surmise has occurred), and also 'Long Burning Lamp Oil'. Signal lamps had to be kept lit whenever trains might run during the hours of darkness, some minor branch lines with trains during daylight hours only would have lamps lit for part of the year only. Otherwise it was the duty of the lamp-man to clean and refill signal lamps on a daily basis; that is until 7-day long-burning lamps were introduced at some locations. Even so if a lamp went out it was the signalman's responsibility to ensure it was re-lit and if the lampman could not be summoned then the task fell on to the actual signalman. It may also be interesting to note the branding on the barrels. The overhead drums perhaps containing supplies in bulk which was then used to charge the relevant barrel. In 1934 and according to Swindon Engineering Society paper No 200, GWR oil consumption was; 590,000 gallons of blended engine oil, 225,000 gallons of cylinder oil, 230,000 gallons of carriage and wagon oil, 380 tons of wagon axle grease *HG1 008 (72037)*

The unseen Great Western

The unseen Great Western

Opposite top: An example of a station on a rural branch line, although one sadly no longer with us. This is Christow on the Exeter Railway - Exeter to Heathfield line looking towards Exeter and just two weeks before passenger closure. The line here had opened with due ceremony on 30 June 1903, public services beginning the following day. Between 1903 and 1911 Christow was the only passing place and then only for either two goods, or one goods and one passenger train, this was altered to two trains of any type from 1943. The Teign Valley line had access to a number of stone quarries which provided road stone for improving and expanding the road network. Unfortunately this traffic was also the indirect cause of the line's decline as buses now used the new roads built with that same road stone. Notwithstanding the decline, the GWR placed camping coaches at some of the stations on the route whilst an additional halt had opened at Chudleigh Knighton in 1924. It was left to British Railways to deal the fatal blow and passenger services were withdrawn in June 1958 followed by freight in May 1961. Meanwhile at the Exeter end of the line a section was saved as an oil terminal, timber yard and scrap yard. Resurrection for the complete line as a through route was considered in 2014 in consequence of the coastal damage at Dawlish. Unfortunately some sections had already been swallowed up by the A38 whilst localised flooding had been the cause of freight services being withdrawn in 1961. Should further damage occur at Dawlish reopening may perhaps again be considered. *HSC4 002 (58002)*

Opposite bottom: WW1 scene at Codford on the line between Salisbury and Westbury. The men are almost certainly volunteers (lambs to the slaughter) awaiting transport. Aside from local men who willingly took the 'King's shilling', Codford, on the fringe of the Salisbury Plain training area, was also the junction of a 2¾ mile standard gauge military railway leading to a military camp. The military line was taken over by the GWR in 1918 but closed in 1922. The station however remained in use to serve the small village of the same name but closed to passengers in 1955, goods surviving until 1963. After this it was only the signal box and that too was closed when the adjacent level crossing was automated in June 1982. Today trains still run over the section of line with only one of the intermediate stations, Warminster and one Halt at Dilton Marsh on the outskirts of Westbury remaining open. *HSC$ 078 (58078)*

Above: Banbury with its overall roof, There were two stations here, one GWR and one LNWR (later LMS) at Merton Street. Both displayed the same name until 1938 when the much larger GWR station was given the nomenclature Banbury General. Post 1961 when Merton Street had closed it reverted to simply Banbury. The GWR station was not in good condition prior to WW2 and a rebuilding had been planned. The eccentricities of war however forestalled those plans and the new station without an overall roof and based on concrete and glass opened in 1958; the overall roof being deemed unsafe had been removed five years earlier. *HSB6 005 (101690)*

The unseen Great Western

Above: Three 57xx pannier tanks, two may well be 5778 and 7743, battling winter conditions at Craig-y-nos on the former Neath & Brecon line. No date but a reasonable assumption would be the severe winter of 1947. The rear engine at least is in steam, and has rescued the other two 'dead': residue heat from the top of the boiler and firebox would otherwise never have permitted such an accumulation of snow. It was a few miles east of here at Dowlais Top that the GWR experimented with jet engines to blast snow away from the rails. It was a short lived experiment as the jet engines mainly blasted sections of frozen snow and ballast combined sending these in all direction rather than the intended melting that had been hoped for. In some places where the snow was so deep as to form a tunnel through which the jet engines would attempt to clear, the result was a clear risk of asphyxiation to the crews. The propelling engine could also find itself being pushed backwards such was the force of the jets. *HLO5E 078 (57868)*

Opposite top: Open cab pannier No 646 at Birkenhead on 2 June 1932. The driver may well be making up his log whilst the fireman appears to have an almost respectful look about him. The engine is a member of the '645' class. A total of 36 engines of this type were built in batches at Wolverhampton between 1872/73. They were very similar to the (first) '1501' type all of which were originally provided with saddle tanks with most later receiving pannier tanks as seen. No 656 had a life of 62 years from May 1872 to July 1934 and was fitted with pannier tanks and superheated in October 1923. The first vehicle of the train is No 3596, a 70' Bars 2 Toplight Brake third of 1913. *HLO5E (57846)*

Opposite bottom: This time it is No 1872, an '850' class pannier tank dating from 1890 and also originally fitted with saddle tanks. It was rebuilt in the form seen in September 1915. Again we have the proud looking driver and respectful fireman. No date for the image and no official location but I think we may say it is at Birmingham Snow Hill. We might also assume this a local headcode to the Birmingham area and where No 1872 was acting as station pilot. All of the 850 class were built at Wolverhampton, this particular engine in 1890. The type might be seen literally all over the system, their diminutive size gaining them the reputation of being the equivalent of the Southern 'Terrier' class. The comparison in size between the engine and coach proves the point. The disc '9' on the bunker probably referred to this particular duty. *HLSE 073 (98690)*

The unseen Great Western

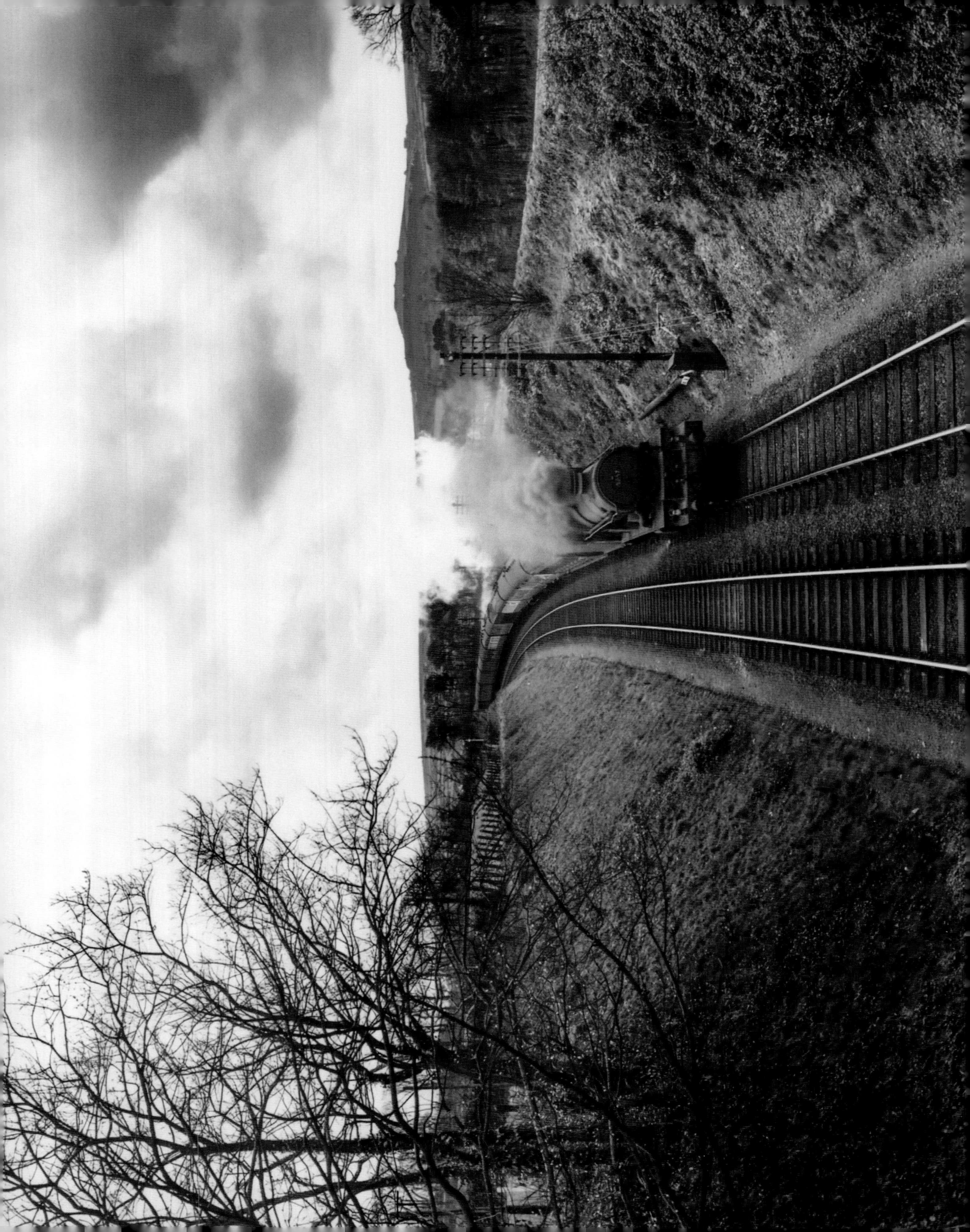

The unseen Great Western

Opposite: Castle class 4-6-0 No 5057 *Earl Waldegrave* on a mixed train and with Class 'A'. headcode. No 5057 entered service from Swindon in June 1936 carrying the name *Penrice Castle*. Allocated initially to Newton Abbot it remained in the west country at Exeter and Laira until the start of the BR era and thereafter moved around to locations such as Old Oak Common, Bristol Bath Road and even Banbury for a short time in 1959/60. The engine would be withdrawn from Old Oak in March 1960. The location is not given but it appears No 5057 has just had coal added to the fire perhaps in anticipation of an approaching gradient. From the stock behind the tender the first vehicle appears to be a Collett 10-compartment third, followed by may be a Mk 1. The remaining pair of passenger vehicles are too indistinct to afford judgement. The remaining seven vehicles are all vans, respectively of GWR LMS and LNER origin; clearly then a train dedicated far more to parcels than passengers - a newspaper service perhaps? *HL13C 070 (65439)*

Above: This time it is No 5005 *Manorbier Castle* that is seen at the intermediate block signals (electrically operated) at Basildon (near Pangbourne) with the famed 'Bristolian' service, sometime in 1935. No 5005 was a standard Castle class engine built at Swindon in June 1927 and remained in service until February 1960. The mid-1930s witnessed the height of the streamlining craze, unexpectedly led by the GWR whose 'semi' treatment actually pre-dated the fully-clad LNER A4s and LMS Coronation Scots. Folklore has it that at Swindon Mr Collett was persuaded by higher authority to take part and reluctantly modified one Castle class and one King class engine based, it is rumoured, on lumps of plasticine stuck to models on his desk. Whatever the origins, the results are seen here with fairings on the front footplate, against the cylinders, behind the chimney and safety valve, between the boiler and firebox, and on the running plate so as to combine the splashers and further back above the coal space (an elongated nameplate was now necessary for the engine). A wedge-fronted cab was also fitted; this was really the only good idea as it reduced reflection at night) whilst the most noticeable modification was the hemispherical protrusion attached to the smokebox door. It would be hard to say even the most ardent admirer of the GWR could approve of the results. Even so in this form the engine ran in traffic although not without some difficulty as access for lubrication and maintenance was now reduced and there was a tendency for some bearings to run hot. Gradually then the various streamlined items were removed and No 5005 reverted to type. Contemporary magazines have it that 100mph was achieved in this form but this was not enough to convince Swindon the modifications were worthwhile. *HL13F 014 (65511)*

The unseen Great Western

Above: No 5922 *Caxton Hall* approaching Chippenham from the Bath direction with an Up Class A passenger service. Chippenham, like Bath, had a centre road not used for through running, the train seen here will go around the north side of the island platform whereas through down services used the main platform. Access between the main and island was by a footbridge. No 5922 was built at Swindon in May 1933 and had a life of just over 30 years ending its days at Oxford in January 1964. It is seen here in none too clean BR livery and with the first BR emblem on the tender. The first coach at least is a Collett brake third but identification of the rest of the train cannot be confirmed. Between the two running lines is a headshunt which is occupied by a pannier tank and a 6-wheel syphon branded to the sausage and bacon producer 'Harris Calne'. The photograph was likely taken from Chippenham West signal box which stood immediately behind the photographer on the left hand side of the line and worked to Chippenham East, and west to Thingley Junction; the latter the point where the line through Melksham and Westbury diverged. *HL12B 024 (74105)*

Opposite top: 'Aberdare' 2-6-0 No 2653 on a long train of ASMO vans loaded with commercial vehicles and chassis at Morris Cowley (Oxford) on Friday 4 November 1930 at the start of their 370 mile journey to Glasgow. The vehicles and equipment were to be on show at the Scottish Motor Show which opened its doors on Monday 7 November 1930 and ran until 15 November. The train is facing toward Oxford and so will likely be transferred to LMS metals around Birmingham. There appear to be 15 wagons in the train plus the brake van and it will be noted there is another engine at the far end, of the 4-4-0 type. The ASMO type van has end loading doors so vehicles could be driven on to and in theory the complete length of the train. The term 'ASMO' is believed to mean 'assembled motor cars' although in this the vehicles were drivable chassis. Engine No 2653 had entered service in 1901 and lasted until 1934; its demise coming about owing to the conversion of a 2-8-0T to the 2-8-2T so affording a more modern freight engine suitable for medium distance work. It is seen here with a former ROD tender; a case of using resources wisely with the original tender from this engine being used on a more modern engine. *HL18A 130 (90359)*

Opposite bottom: Another slightly longer motor car train, 18 vans, also near Morris Cowley on 14 November 1933. In charge is Churchward 2-6-0 No 6340 dating from 1921 and which had a life a few days short of 41 years being withdrawn in July 1962. The train has also been branded, almost certainly with Morris adverts and ironic to think that like the granite seen sourced from the Teign Valley line earlier, the contents of this train would be in the vanguard of the move towards independent road transport and with it the demise of so many stations, branch lines and cross country routes. Morris Cowley was on the route from Oxford Kennington Junction through Thame to Princes Risborough and a useful diversionary line at times. As a through route it has been closed for many decades but cars are still loaded at Morris Cowley today, nowadays usually bound for export through Southampton. There is also occasional talk of reinstating a passenger station here as a local commuter point. *HL18A 129 (90358)*

The unseen Great Western

83

The unseen Great Western

The unseen Great Western

Opposite: Head on view of 4076 *Carmarthen Castle*, one of the early series engines dating from February 1924 and having a working life of one day over 39 years. Originally allocated new to Old Oak Common it would be fair to say this engine travelled widely from Bristol into Wales, north to Wolverhampton, south west to Newton Abbot before finally ending its days at Llanelli. Sometimes such moves were indicative of a shed foreman pleased to see the back of a poor performer but we should not take that as necessarily the true reason and No 4076 may simply have had her 'passport' stamped more times than most. The design of the first series of Castle class engines is all too clear, with the curved sides to the valve chests for the middle cylinders. Red loco lamps also indicate this is a 1930s view. The photograph was taken not at Swindon Works but instead at Swindon shed which lay to the east of the Gloucester line and north of the stores building referred to earlier. With no ash deposits on the front framing it may be No 4074 has been sent new to Swindon shed for 'running-in'. *HL13A 047 (65082)*

Above: A most interesting view and as seen a little bit more than a standard 4,000 gallon tender, as originally coupled to several of the late 1926 batch of Castles, as well as all the 1927/28 Kings. This one has already been modified with proper coal doors rather than the early arrangement of coal boards dropped in between two angles riveted back to back on each side of the bunker. Top left is an accurately calibrated tank level gauge rather than the more simple standard tank gauge. It is not possible to read the printing on the face of the box on the right (as viewed) but the gauge is a standard GWR vacuum gauge as used with the brake system. Likely this apparatus was for the purpose of accurately measuring water flow rates to the exhaust injector which was situated on that side of the engine along with its supply pipework on the same side of the tender. The gadget with the handle at the front of the water gauge is undoubtedly a hand driven gear pump although its purpose is not known. There are also additional feed water cocks on each side (these have angled handles) and are probably connected to watercocks in the additional copper pipes which have been tee-ed into the standard water feed pipes on each side. Possibly an alternative supply for each injector for monitored use during test periods. Using as source referencing the Andrew Cole database, this is almost certainly tender No 2384 and the second of the 4,000 gallon tenders to be built, it cost £1.037 when new. Tender 2384 is shown as being allocated to experimental work for 80 days from 9 January 1931. This corresponds exactly with experimental work conducted with No 6005 *King George II* between the same dates. We should also note the tender has a full complement of hand picked best Welsh coal which has obviously been carefully loaded and stacked by Swindon's 'formation dry-coal walling team'. The care with which this coal of optimum size has been selected and loaded has to be seen as a clear indication that this tender has been prepared for coupling to an engine intended for road testing. Note too the temporary chain coupling and 'dumb buffer' on one side; both indicative the tender has been shunted into this position. Despite referencing all known published sources, it has not been possible to ascertain the nature of these tests. O S Nock in his 'Stars, Castles, & Kings Part 2' has an undated contemporary GWR era image of No 6005 complete with front indicator shelter but without further explanation. The works turntable survives in 2025 albeit fenced off. Otherwise everything else around has been removed/demolished to be replaced by car parking and open land. *HL14A 062 (53327)*

The unseen Great Western

Opposite: Two views of 'Saint' class 4-6-0 No 2935 *Caynham Court* inside Swindon works and almost certainly in 1931 at the time the engine was modified with Lentz style rotary cam poppet valve gear. No 2935 dated from 1910 and had been part of the first batch of 10 'Court' class engines within the 'Saint' series. They were all built as 2-cylinder engines and for many years performed turn and turn about on the same duties as the 4-cylinder 'Star' class. As built No 2935 had piston valves, but in 1931 was modified as seen. The new Lentz style gear allowed the driver to make nine different adjustments of the reversing gear from 10% to 85% as required. Unfortunately this cab view does not show enough of the controls and in particular the now different reversing gear. As seen the front bogie remains to be fitted but otherwise the engine appears complete. This experiment had been completed in the time when C B Collett was in charge at Swindon, a man not known to favour the innovative attitude. It appears most likely that following experimentation on two former 4-cylinder engines of the Great Southern (Railways) in Ireland, the GWR were induced to modify No 2935 on the basis the outside Associated Locomotive Equipment Co, (who produced the bevel gears and other components) paid the bill. A report in the 'Railway Gazette' for 1931 described the modifications to No 2935 but which did have the disadvantage of adding two tons to the overall weight of the engine. This was contrary to the manufacturers claims that the gear was lighter than a conventional gear but is explained by the necessary addition of a gearbox, shafts and universal joints etc. None of these were in normal use or manufacture at Swindon. Following completion, No 2935 undertook road trials for about 18 months. The paperwork appertaining to the actual tests has not been located but Mr S O Ell did produce a paper in 1933 surmising some of the findings. Broadly speaking it was found the expected benefits were not realised whilst performance was hampered by leakage of steam past the valves; notwithstanding various modifications to these having been made. Serious testing of all sorts appears to have ceased around 1933. Possibly the reason No 2935 did not come up to expectations was that there was little difference compared to what had already been a good 2-cylinder engine. But that was not quite the end of the story, for whilst No 2935 retained the same valve and equipment until withdrawal in 1948, it is suggested that Hawksworth was not convinced the potential for Poppet valves had been fully explored with private discussions being held between parties that did not involve Collett. Indeed Hawksworth is said to have considered fitting one of his 'County' class 4-6-0s with similar equipment on a post-war railway. It is interesting also to consider that this was a 20 year old 'Saint' rather than a more modern 'Hall' that was used for the trials. Was Collett concerned one of his designs might be held up to scrutiny should it fail and this is why he used the older (Churchward) design? Might he even have seen the proposed drawings and 'modified' these so as it 'was designed to fail'? Poppet and similar valves would have a resurgence on steam elsewhere in the world later although in the UK only one engine was so fitted, No 71000 *Duke of Gloucester*. In the background is a 'Westernised' 'ROD', No 3039. *Top: HL10B 003 (74660), bottom HL10B 004 (74661)*

Above: Standard bogie or truck wheelset with the axleboxes and ATC gear assembled to demonstrate the arrangement for the benefit of the official photographer - and NOT then a pony truck; reported as taken on 27 September 1947. The wheelset would be capable of being fitted to a wide variety of engines. It would only be used in an engine built from around 1932 onwards when the location of the ATC receiver moved from under the back of the cab to the front of tender engines. That date similarly confirmed as the axleboxes fitted to it are of 1932 pattern. Hence this could be for an engine from the previous 15 years or even a new one. The wheels appear to be standard 3' 2" diameter with full width tyres, which rules out the 'Hall' class which had 3'0" diameter wheels with narrower 5" wide tyres. Otherwise this could be put under a Castle, or any later 28xx, 41/51xx, 61xx, etc. The wheelset and components appear newly overhauled, the ATC shoe insert is unworn, the corks in the underkeeps look new, and the axlebox guides and tyres look untroubled by use. As to why the photo was taken is the mystery. There is a similar view but taken from the rear, that appeared in the October 1947 GWR booklet referencing the high speed test between London and Reading for ATC and four aspect signalling but that may not be the explanation. Tyre widths varied slightly according to engine type. *HS3 033 (87886)*

The unseen Great Western

Above: One of a very limited number of GWR tank engines to bear names. This is No 17 *Cyclops* built in 1901 using as the basis the '850' class design but with the frames extended backwards on to which was mounted a steam crane in lieu of the conventional coal bunker. The wheel arrangement was therefore an 0-6-4T. Although not a true member of the '850' type, this and sister engine No 18 *Steropes* were the first '850' type to have pannier tanks; saddle tanks would clearly have been pointless. The crane was also steam powered and could lift a maximum of around 10 tons dependent upon the radius from the engine and the number of chains being used. Examination of crane and locomotive reveals the absence of the out-riggers that are usual with conventional railway cranes. These 0-6-4Ts were required to operate within confined spaces, for example in workshops or against platform faces, where such equipment would have been impracticable while also impeding mobility. Stability was ensured by the rearward cantilever of the crane's base frame on which was mounted a heavy counterweight thereby allowing a load capacity of maximum six tons on an 18' radius and nine tons on a 12' radius. For travelling purposes, the jib was stowed forward across the cab and rested on a transverse bracket roughly in line with the firebox front. To respect the loading gauge, a pair of domeless round-topped boilers were specially designed for Nos 17 and 18. Forward vision which otherwise would be limited by the jib, was improved by the unusual rectangular cab spectacle plates; this style was also used with at least one member of 0-6-0PT Class 1076. In 1921, No. 16 *Hercules* was added to essentially the same format except that it carried a domeless Belpaire boiler from new. The entire ensemble was an ingenious solution to the embedded challenge of creating an effective crane-engine. They were mainly associated with Swindon although one was allocated to Stafford Road for several years until the run-down of Wolverhampton works. They were also occasionally used at Reading signal works. Installation of modern electrically-powered internal cranes at Swindon rendered the trio redundant, leading to their withdrawal in 1936. *HL19B 050 (94165)*

Opposite top: A wonderful rear three-quarter image of the original No 4082 *Windsor Castle* not quite in the condition seen by their majesties King George V and Queen Mary when they visited Swindon in April 1924. No 4082 was brand new at the time and as has been well reported, was driven by the King from the works to the station at the end of the royal visit. Of interest here is the commemorative plate attached to the cabside above the number. Notice too the fireman's shovel neatly stowed above the handrail on this 3,500 gallon tender the contents of which are perhaps not 'best Welsh'. On the right is what may be a Dean Goods; whatever it is definitely in ex-works condition. In the distance are the frames less boiler of a ROD; the number '3001' chalked on the frames. No 4082 exchanged identities with No 7013 in February 1952 so that the same named engine might haul the funeral train of the later monarch King George VI. (Research by Bob Meanley within the STEAM archives has shown that following the funeral Nos 4082 and 7013 resumed their original identities but only for a matter of days as instructions were received from the General Manager at Paddington, Mr K W C Grand that the temporary change was to be made permanent) *HL13A 112 (65147)*

Opposite bottom: A wet day at Swindon in early 1951. No 7916 *Mobberley Hall* is being fitted for tests which took place both stationary on the works test plant and also on the road with the dynamometer car. During the static tests the engine was run at the equivalent of 40 mph with full open regulator and 24% cut off. The boiler pressure was recorded at 205lbs psi with the difference in pressure between the boiler and steam chest a remarkably small 2-3 lbs. Proof that the steam passages on the class were well designed. The 'Eagle' comic produced a coloured cutaway and drawing of the test train complete with dynamometer car in its edition of 11 April 1953. No 7916 was built at Swindon in BR days and completed on 6 April 1950. It spent most of its life in the south-west before moving to Cardiff East Dock in 1964 and from where it was withdrawn, week ending 26 December that year. Alongside is No 6803 *Bucklebury Grange*. This engine entered service on 5 September 1936 and lasted until 25 September 1965. *HL12D 014 (62121)*

89

The unseen Great Western

Above; No 1000 *County of Middlesex* on controlled road testing passing Swindon works on the down main line. (Another engine involved in what were protracted steaming trails was No 1009.) There were several test runs involving the 4-6-0 County class, the aim being to achieve the optimum performance as front end draughting was not quite as had been intended when first built. The class eclipsed the existing Castle type with an increased boiler pressure of 280psi (a Castle was 225psi and even a King only had 250psi.) Some drivers seeing the increase in pressure therefore imagined them to be 'super' engines and attempted to drive them accordingly and then quickly ran out of steam. Mr Hawksworth is said to have been influenced by Bulleid on the Southern when it came to boiler pressures allegedly commenting along the lines of '...if it is good enough for him..... . '. The outcome of the various trials concluded a revised double chimney design was required and this was subsequently incorporated. The train seen would have comprised selected stock made up to the required weight. Then the engine and dynamometer car would be attached. On the front of the engine is a wooden shelter where observers would take further measurements. The circular wooden apertures supposed to assist the driver in looking forward. It is believed this working had started at Reading and would continue through to Stoke Gifford (more recently known as Bristol Parkway). Boiler pressure was later lowered to 250 psi reducing tractive effort but more significantly reducing boiler maintenance costs which hitherto had been among the highest in the BR fleet. Notice the ATC ramp just ahead of the engine, deliberately placed here as there are stop and distant signals on the same post. On the left hand side a number of works staff have come out to watch the procession. The Hawksworth County class engine ended up mainly working trains in Cornwall and on the north-west route. However, a few were deliberately put on London-Bristol workings so the designer might look out of his office window and observe the passage of one of 'his' engines. *HL10C 138 (85531)*

Opposite top: A remarkable image, undoubtedly one of a series but of which this is the only one located. It shows the result of 'slipping trails' involving an 'Austerity' engine - presumably a 2-8-0 on 7 January 1952. The rail has been completely distorted and whilst this is clearly an extreme example, it is certainly true that an unexpected slip could indeed cause an immediate problem unless dealt with at once. Driver's would talk of a brief second when the sound changed before the commencement of the actual slip which is where the skill of the man with his handle on the regulator came in; to close the throttle and at the same time apply sand. It is believed the location is Swindon with the same ringed arm signal as is seen in the lower view. The identity of the engine and what damage it too may have suffered is unknown. *HA1 059 (64949)*

Opposite bottom: It is almost tempting to suggest this was a similar experiment but this time with 2884 class 2-8-0 No 3851 and now with the location clearly outside the works. No details as to the date but the engine does appear in good external (black) livery and with the first BR crest on the tender. Clearly the driver has been instructed to act in a particular way whilst no less than four observers are watching on the offside as well as others not so likely to be involved by the shed wall. From the column of smoke the noise would have been terrific. *(As a Great Western devotee might we comment was it not against the law for a mighty 28xx to slip....?) HL08E 085 (90706)*

The unseen Great Western

Austerity Engine Slipping Trials — Crippled Rail

A3/302
7/1/52

The unseen Great Western

Driver Arthur Selwood of Swindon seen on the footplate of Castle class 4-6-0 No 5056 *Earl of Powis* posed on the Swindon works turntable, 21 April 1944. Built at Swindon and completed on 17 June 1936, this engine originally carried the name *Ogmore Castle* but a name change took place in September 1937. (The name *Ogmore Castle* was carried briefly by No 7007 – the last GWR-built class member – and finally by BR-built No 7035.) The engine seen here is fresh from overhaul and carries unlined war-time livery. The cabside window has also been replaced with plain sheeting; it would be reinstated later. Even so still a fine machine which had its early life working from Old Oak Common before spending its final 18 months between Cardiff East Dock, Hereford, and even a short spell on the London Midland Region from where it was withdrawn on 21 November 1964. On the footplate Driver Arthur John Barnes Selwood was born on 4 July 1883 and entered GWR service as a cleaner at Swindon on 3 July 1899, his rate of pay was then 5/6d per day. He progressed through the usual ranks of shunter/fireman, third, second and first class fireman similarly gaining experience working at Llantrisant and Severn Tunnel Junction in the process. He was back at Swindon on 30 June 1920, a note on his personal papers adding 'own engine' unfortunately without elaboration. The same paperwork records no less than six injuries received at work fortunately all various cuts and bruises but still some sufficient to warrant time off work for recovery. He was commended in July 1927 for 'Being well on the alert and promptly blowing the brake whistle when noticing that the 7.35am goods ex Southall, was being set back towards the junction at the same time as the 12.45pm passenger from Swindon to Bath was leaving Wootton Bassett station with all signals off. The driver of the 7.35am upon hearing the brake whistle, brought his train to a stand which no doubt averted a serious collision.' Mr Selwood had the honour of being mayor of Swindon in 1942 and retired from the GWR on 24 November some nine months before his 65th birthday; slightly surprisingly he had a further medical examination when he was retired, on 30 December 1945, Dr Bennett noting 'Capable of carrying out normal duties had he not retired.' His retirement was not as long as might have been hoped for he died just five years later on 10 September 1950. *HS2 079 (81764).*

The unseen Great Western

Another locoman posed, this is driver Chaplain of Wolverhampton, alongside an unknown 'Castle' on 3 September 1945. William Henry Chaplain was born on 7 August 1881 and joined the GWR at Wolverhampton on 22 July 1897 as an engine cleaner. He became a first class fireman on 8 July 1901 the same year he was cautioned for his light engine colliding with the rear of a goods train. He was promoted driver at Didcot in 1912 and again unfortunately the same year he drove engine No 659 into the stop blocks causing 'considerable damage'. Following this move to Didcot his career took him back to Stafford Road. William had a total of 48 years on the GWR finishing work on 29 September 1945, a note in the records simply stating 'Retired - Compulsory - Age'. This view was taken in the month he ceased work. At the time he started on the footplate he would have been with men to whom the broad gauge was familiar and yet they and indeed his stories have sadly likely been lost forever. *HS2 003 (53872)*

The unseen Great Western

Above: On today's railway of fixed formation workings, it is almost 100% predictable what will appear as the next train at a station or on a particular stretch of line. True, there are some variations, empty stock workings, some limited freight but this will depend very much on the location and above at Newton Abbot traffic is now almost 100% passenger. Not so a few decades past such as this wonderful portrait of 47xx No 4703 leaving with a mixed parcels / empty stock working. The first vehicle is a 'Cordon' gas tank wagon, no doubt on the way back to Swindon for re-gassing after which it will be returned to perhaps even Newton Abbot or maybe Plymouth where the gas will be used to refill the cooking tanks on restaurant cars. Speaking of the latter the next vehicle is a catering coach which appears to be either a Kitchen Car or possibly a Diagram H53 Gangwayed Buffet Third (Sunshine stock), perhaps an unbalanced working; notice the longitudinal gas tanks under the coach. Notwithstanding the presence of flammable gas when a 'Cordon' was full, there does not seem to have been any particular instructions about where in a train such a vehicle might be formed. Behind the kitchen car are a pair of Southern Region utility vans, the remainder of the train indecipherable. No 4703 was one of a class of nine engines intended for fast fitted freight and parcels workings. This particular example was built in 1922 and lasted until 1964. None survived into preservation although there is a 'new-build' making progress in the 21st century. *HL08E 046 (60782)*

Opposite: A wonderful view of the docks at Birkenhead and from the wording on the warehouse there is no doubt as to ownership. The variety of wagons also gives a good indication as to the type of traffic handled - basically anything and everything. Many of the wagons are of course of Great Western origin but it is interesting to note there are also a number of 'common user' vehicles. Note too the pair of travelling cranes; hand operated of course and usually requiring four men per crane. A docks at Birkenhead had first been proposed in 1820 but nothing came of the idea. It was not until 1847 that the first docks were in operation. Much in the way of securing a satisfactory connection to the railway network followed and it was not until 1854 when the GWR had taken over the Shrewsbury & Chester and Shrewsbury & Birmingham railways (much to the chagrin of the LNWR) that access was now possible via Chester. Relations between the GWR and LNWR did eventually thaw so that six years later a joint company was established which remained independent until nationalisation in 1948. In the mid 19th century the percentage of trade dealt with at Birkenhead was just 2% of that of the Mersey and comprised exports of Welsh coal and imports of timber and guano. The latter used as fertilizer. Available waterfront though meant several businesses were established including ship building. The fact the docks developed slowly was in effect an advantage as railway lines could be laid to suit the positioning of docks and warehouses and in this respect it was better laid out than across the river at Liverpool. As years passed so traffic developed that included grain, frozen meat, in fact most goods. Some colliery owners from Wales preferring Birkenhead as the hoppers were lower and there was less breakage of the soft Welsh coal. Several named docks existed at Birkenhead including Alfred, Bidstone, Egerton, Vittoria and Wallasey. Three remain although only two, the Alfred Dock which is the entrance to the River Mersey, and the Vittoria Dock remain operational. There are also a number of quays on which are towers/ facilities for grain, mortar, petroleum and oil. Birkenhead Docks was one of three locations on the GWR where the shunting engines regularly carried bells to supplement the engine whistle; the other two being at Cardiff Docks and on the Weymouth Harbour Tramway. (A bell was of course also carried on No 6000 *King George V* following its visit to the USA.) *HS86 002 (101681)*

The unseen Great Western

Factory

Tare 4.19

G.W

A wonderful posed view from the early 'Is it Safe?' movement recorded by the official photographer. The question being asked related to the safe use of the pinch bar; although this particular wagon will not be going anywhere fast as it will be noted the brake is on! Dealing first with the wagon, this is an open used for internal 'Factory' use the 'X' painted on the side door confirming the fact. The tare weight is just 4ton 19 cwt and it will be noted to have grease axleboxes and wooden brake shoes. Use of a pinch bar to move a single wagon was a commonplace if risky occupation. The bar having first to be wedged between the wheel and the rail and pressure then applied to start movement - not always successful and sometimes requiring the efforts of more than one man. As with the horse we saw attempting to move the wagon earlier, much also depended upon the ground conditions whilst even the slightest gradient could either make matters impossible or create the conditions for a runaway within the yard. The coach behind would appear to be fresh out of shops and is in crimson livery. It is Clerestory Diagram C10 All Third No 3021 from Lot 823, completed 26 December 1896 and notionally seating 80 passengers in ten compartments. This type was numerous, and built under at least nineteen lots between 1894 and 1902. Many survived into BR service including four of Lot 823 [but excluding No 3021]. *HS2 003 (53872)*

Bristol No 2 tunnel and the nearby signal box in mixed gauge days. St Anne's Park station was opened on the site between the signal box and the tunnel in 1898. At the bottom of the image is a narrow gauge siding, still with baulk road; opposite this on the far side of the track is one of the sandstone boulders excavated from No 1 tunnel which is just behind the photographer. The signal box here had a short life, comprising just 14 levers it opened in 1890 and closed from 4 February 1909. *HS3 057 (69301)*

Wolvercot sidings on the north side of Oxford. The official name should have been 'Wolvercote' as per the nearby village but the name was altered to avoid confusion with a nearby halt of the same name on the LNWR line from Oxford to Cambridge. For its part the GWR added to the confusion with their own Wolvercot Halt on the line between Oxford and Kidlington which was open from 1908 to 1916. There was a definite railway connection with those responsible for the GWR's Wolvercot Halt, as the stopping place had once been under the control of Miss Margaret Elsden, the sister of Mr A H Elsden who became stationmaster at Birmingham Snow Hill. To confirm company loyalty even further, Margaret Elsden later married Frank Buckingham who became stationmaster at Oxford. The view shows the level crossing and wicket gate running across the three running lines and two sidings. These were controlled by a signal box located just behind the photographer on the left hand side. Being a public crossing it is possible the public right of way may well have been obstructed for some time should a goods train be occupying either the down siding or down goods loop. A note attached to the image indicates the view was taken in connection with an accident of some sort but without further details. The view to the right is towards Oxford. *HS3 082 (61327)*

The unseen Great Western

The unseen Great Western

Opposite top: Mention has been made already of the army of clerical staff necessary to run the railway. Here we have an example of three at work in an unknown booking office. Clerks were part of the traffic department and graded according to their role and skills. An entry level post might be as a junior clerk, where after having gained experience in the working of the booking and goods offices and passed examinations in station accountancy and signalling (train working) plus of course the rule book, might well progress through the grades to become deputy to the station master. (Station masters were similarly graded according to their station's importance and traffic handled. Each day the numbers of each printed ticket issued for specific destinations, its class, and status (single / return etc.) was noted and at the end of the day the last number issued recorded. Added to this were 'blank cards'; tickets issued to a destination for which there was no regular ticket. The value of all of these was then added up in a ledger and which should tally with the amount taken in the till. The details and amounts were then sent to the main station in the division and the monies banked. (Small stations would use a cash bag and a travelling safe in the guard's van of the branch train.) On regular occasions the auditors would visit from Paddington when all from the station master down would hope all would be found to be satisfactory. HS3A 150 (101739)

Opposite bottom: Whilst clearly run as a business for the benefit of the shareholders, the GWR was also very much a family railway, employing generation after generation of the same family; indeed if a relative already worked for the railway, a new applicant for a post was at a distinct advantage. As well as affording employment, at several locations the GWR also provided housing, this image depicting a block of six terraced houses for staff at Severn Tunnel junction. There is also an individual property, its status meaning this would be for a station / yard master or similar role. In the early years of the 20th century, staff such as a station master of sufficient grade might even have a housemaid engaged on their behalf. Should railway accommodation not be available then it was practice to rent a house within close proximity to the railway. Today many of these railway built houses survive, we have mentioned surviving station master's houses previously, but blocks of dwellings still stand, their occupants very often unaware of the buildings' humble roots. *HS2 087 (85014)*

Above: We could not exclude a view in what might well be referred to as the 'pastoral' series, examples of which were to be found in waiting rooms and especially on the compartment partitions of carriages. This particular example is the River Wye, from the boundary of Gloucestershire and Monmouthshire and with the ruined Tintern Abbey on the Welsh bank. Literally thousands of these images were taken by the official photographers intended to promote travel to areas served by the GWR. This in addition to their other work relating to the products of Swindon as well as infrastructure changes. Most pastoral scenes were just this, pastoral, but it will be noted the photographer had no choice but to include a part of the Wye Valley line crossing the river in the background. Just a pity perhaps there was no train! *PC3 065 (72561)*

The unseen Great Western

Above: 'Modern' traction? Well certainly ground breaking for the time as this is the short lived Metropolitan Vickers gas turbine electric No 18100 seen on the sea wall at Dawlish and running east after a trip to Plymouth. This was the second of the gas turbine electrics delivered to the Western Region although both had been ordered prior to nationalisation. No 18000 had been built in Switzerland and was designed to burn heavy oil. It was reasonably successful although the life of the heat exchanger was limited. So far as power output was concerned it was rated at 2,000hp and was thus the equivalent of a 'King'. It managed to survive in service until 1960 although towards the end was operating on three instead of four traction motors meaning its output was 1,500hp. After withdrawal it ran as a test machine in Europe before the shell was returned to the UK where it resides at Didcot. No 18100 was a far more powerful machine, 3,000hp at a time when 'turbines' were seen as the power of the future for everything from aircraft, to boats, trains, cars and even experimentally on a motorcycle and bicycle. Unfortunately a gas turbine is only ever efficient if working at full power and there simply were not the trains and schedules to allow this to happen in the steam era. Also No 18100 was designed to burn ordinary diesel fuel (and an awful lot of it). British Railways simply could not find a means of operating No 18100 efficiently and in consequence it was withdrawn from WR use in 1952 (not 1958 as is stated by so many sources), and with the intention that it too be converted to burn heavy oil as per No 18000. The conversion was attempted by the manufacturer in Manchester, we say 'attempted' as after six years it was admitted this was simply not viable, consequently No 18100 was withdrawn from capital BR stock in 1958. After this it had a radical conversion into a prototype for the forthcoming Manchester 25kV electrification, the cab controls replicating those of the soon to be delivered new AC electric engines. It was also given the number E1000, later E2001. Unfortunately it would not survive a second life, was withdrawn and scrapped in 1972. *H03 037 (62158)*

Opposite top: Cab view of one of the new North British 2,000hp diesel hydraulic locomotives D600 at Swindon on 28 January 1958. The Western Region had opted for hydraulic transmission for their modernised locomotive fleet for the very simple reason of the problems encountered with the pair of gas-turbine electrics. To be fair this was partly their fault as the engines were serviced amidst the dirt and grime of a steam shed environment and which led to issues with traction motors. (The LMR and SR experienced the same with their early fleets of diesel-electrics but were wise enough to recognise a clean environment was necessary. The WR however whilst identifying new depots were needed persisted with the hydraulic policy and although it did have some advantages, ultimately BR decided on a standard diesel electric traction fleet and diesel-hydraulic engines were prematurely withdrawn. *HD3 089 (86733)*

Opposite bottom: From the cab of a new 'Hymek' D7000, we see small pannier tank No 1604 at Swindon. To some on the operational side of the Western Region, the Hymek type was the most versatile of the hydraulics but as mentioned above was also to become a victim of the national diesel electric policy and whilst withdrawn in different years, both No 1604 and No D7000 had lives of just 11 years. *HL05E (57858)*

The unseen Great Western

The Story of the Harry Collection

It was a normal day in the autumn of 2011 when the Curatorial department at STEAM took a telephone call regarding a bequest of railway items to the Museum. STEAM receives donations all the time but this particular donation seemed a little different.

The executor of the will (who was not an expert in railway memorabilia) tried to describe some of the items over the phone. He mentioned that there were some very heavy crescent shaped items that were part of the donation. He didn't know what they were, but there were several of them. He also described some heavy rectangular objects that had numbers on them. To the Curatorial team these items sounded very much like the name and number plates from locomotives, but donations of such are very rare, and on this scale unheard of. Photographs were requested in the first instance to establish what was being offered. The team waited with bated breath and it wasn't long before an email came through that confirmed their suspicions. These WERE name and number plates from Great Western locomotives and the executor had found more! They were everywhere –under beds, in wardrobes, in the garage – it was unbelievable.

The team were then invited to come and view the items and find out more about who had donated them. The visit knocked them off their feet. The donation consisted of the largest amount of objects, archive material and photographs ever gifted to the Museum in a single lot. Furthermore many of the items were unique and historically invaluable to the history of the Great Western Railway. Highlights of the collection include nine nameplates, 73 cabside numberplates, four totems, nine signal box nameplates, 15 locomotive tender plates, 40,000 photographs and hundreds of other GWR related objects and archive material.

The collection of material belonged to the late Brian Harry. It was obvious he was a prolific collector, and knew the significance of what he was collecting. Bequeathing his railway collection to STEAM ensured its long term care and future enjoyment by others. The cogs were then put in motion to transfer the objects and material to the Museum.

It was a cold December morning when a team of Museum staff and volunteers headed off to collect the items. The collection took a whole day, and wasn't helped by a snow storm in the afternoon! Two vanloads and four carloads later the items were safely back at STEAM. Many of the objects went straight out on display, and included all of the name and number plates. The paper material, including the photographs, went into the Museum's archive. It took about 5 years to fully catalogue all of the photographs. The images cover a whole range of subject areas including locomotives, carriages, stations, Swindon Works, publicity views and staff. These photographs have become a hugely valuable resource to the Museum and have expanded the team's visual knowledge of the Great Western Railway. The full bequest, including the photographs, is now called the Harry Collection in honour of this amazing donation.

Elaine Arthurs, STEAM – Museum of the GWR, Swindon